What Does This Tell Us?
Social Research Data and How to Come by It

What Does This Tell Us?
Social Research Data and How to Come by It

Mark Taylor

First edition published by Cambridge Academic, The Studio, High Green, Gt. Shelford, Cambridge CB2 5EG.

ISBN 1-903-499-42-9
978-1-903499-42-9

Printed and bound in the United Kingdom by
4edge Ltd, 7a Eldon Way Industrial Estate, Hockley, Essex, SS5 4AD.

Contents

Acknowledgements

I would like to thank the various authors, publishers and organisations who have given their permission to reproduce the illustrations used in this book.

In particular, I would like to thank Clare Bunce of the Cold Spring Harbor Laboratory Archive, Elaine Doak, Head of Special Collections, Truman State University, and Robert Cox, Curator, American Philosophical Society for their information on the Eugenics Record Office archive.

Similar acknowledgements goes to Jane Newton at the British Cartoon Archive (formerly the Centre for the Study of Cartoons and Caricature), University of Kent, Patrick Frazer at the London Topographical Society, and Sue Donnelly at the British Library of Political and Economic Sciences.

Some of the images used in this text have been enhanced using Adobe photo-processing software. The poverty map of London has also been enhanced through photographic processing. Acknowledgement here goes to the labs of Colchester Colour Imaging.

Thanks also goes to the University of Cambridge Library and the Squire Law Library, without whose facilities this book would not have been possible.

1: Research Methods in Context

Introduction

Traditionally, books such as this are intended to provide researchers with the tools with which they can conduct research. In contrast, this text looks at research methods and research design from the perspective of *'This is what can be done'* rather than *'This is how to do it'*. This text will compare and contrasts the various methods, and looks at the implications and issues of using each, rather than offering a 'how to' guide for the use of each particular method. Before doing so, it is necessary to understand the function of the various research methods in the context of the research process, and how this is viewed by the wider society. This is something which will be discussed in the first chapter.

When it comes to (social research) data and the (research) methods used to elicit it, social researchers have tended to be preoccupied with what are arguably 'superficial' descriptors. As far as social researchers are concerned, all that really seems to matter (for the purposes of social research and the conclusions that can be drawn from it), is that the data is valid (shows what it purports to show) and reliable (maintains its meaning over time), and, for analysis purposes, it can be classified as either quantitative (essentially numerical), or qualitative (essentially discursive).

For want of a better analogy, these (*'qualitative'* and *'quantitative'*) descriptors have almost become 'data stereotypes', and this can also lead to stereotypical conclusions being drawn about the data itself. Qualitative data has always had a 'bad press', so to speak (among professional researchers), because it is often difficult to pin down, and it is largely out of the control of the researcher. Quantitative data benefits from its association with mathematics (statistics) and so is perceived to be a superior class of data. This distinction has become less pronounced with the advent of computer programmes which can analyse text for particular content (see below). Content analysis is not the same as discourse analysis, but it does form something of a bridge between the two.

Like the construct being analysed, social research data has itself, often been conceptualised and constructed along artificial parameters. True, in order to apply the correct statistical analyses to a data set it is necessary to be able to identify parameters such as the type of data (nominal, ordinal etc.), the design (independent or repeated measures), the variables in question (dependent or independent), how many types of variables and the analysis required by the research question (correlation or difference) (see Taylor, 2007), but, all too often, data is simply thought of in terms of its role in the different stages of the research process. The 'qualitative' phase of a research project is seen as simply a means to an end, in terms of producing variables which can then be analysed in the quantitative phase of the work. However, 'qualitative' data is in itself a useful source of information.

A substantive part of this book (chapters two to five), details the characteristics and attributes of the various research methods and sources of data, and it does so by way of a typology (detailed below). This typology is designed to show how particular issues are relative to the particular methods and sources themselves and, through this typology, the text will attempt to show how particular issues vary with the method/source chosen, and the implications this has for different research questions. This is illustrated, throughout, with some unusual and entertaining examples.

Also included in this text is a research paper (written by me) which looks at the press' reporting of legal cases, with reference to its utility as a source of information for researchers and legal scholars. The paper sets out to make a contribution to the range of methods and sources available to researchers, and in doing so, it utilises many of the methods discussed in the book: case study, qualitative and quantitative content analysis, published newspaper reports etc.

The Discourse of Research Design and Data Gathering

Although it is a bit of an aside to introduce Foucault into a discussion such as this, it is important to look at what Foucault says about knowledge (what he would call a Discourse[1]) because, on the face of it, his ideas describe much of what this book attempts to impart.

The main tenet of Foucault's philosophy is that society consists of Discourses. A Discourse (in the Foucauldian sense) consists of the knowledge/expertise/terminology pertaining to a certain subject. This is not only limited to scholarly knowledge: the clichés of literary criticism, the rules of a school playground, the jargon of computer experts, the in-jokes at a local bar, and the 'right' designer labels, are all 'Discourses'. People with a greater knowledge of these Discourses have an advantage (power) over those with a lesser knowledge of these Discourses. More than this, the Discourse (or Discourses) determines one's place, and role in society. It also 'locates' the person in that society. Foucault would say that we are 'written' by the Discourse. In other words, the Discourses we know how to use define who and what we are. Our lives are thus governed by Discourse, which we are defined by, and perpetuate in order to maintain our advantage over others. For example, I am excluded from the architectural profession because I do not have their specialist knowledge *(am not able to use the discourse).* If I had such a knowledge then I would find my position in a social hierarchy *(power relations)*, defined by how much architectural knowledge I had and could use *(the Discourse of the architectural profession)*, and would see myself as an architect *(written by the Discourse)*, which would be my chosen profession.

Foucault was (he died prematurely of AIDS in 1979) one of the few (and probably the only one in modern times) philosophers to attempt a 'grand theory'; in other words, a theory that was designed to explain 'everything'. In order to achieve this, he needed a suitable vantage point. The vantage point he chose was (metaphorically) the *edge* of society. From what we know about his private life (in a television documentary screened a few years ago), Foucault spent much of his time exploring the gay scene. This (in theory) enabled him to observe

1. Given that 'discourse' used in the context of Foucault's theory, has quite a specific meaning which is distinct from that used in everyday language, I will use the term 'Discourse', with a capital 'D' to indicate that I am using the word in the Foucaludian sense, and 'discourse' with a lower case 'd' to indicate that it is being used in the every day sense of the word (or as it is used I research when referring to things such as discourse analysis).

'mainstream' society while, at the same time, putting as much distance between himself and (mainstream) society as possible. In many respects, this phrase 'standing on the edge' can be seen as an acknowledgement of the fact that no one can observe and comment on society, *without* being a part of it; if one is on the edge of society then one is still a part of it. It is simply not possible, *not* to be a part of society. (There also has to be a point of reference for the reader.) There is nothing new in this approach; it is simply a version of non-participant observation; Foucault is just redefining what it means to be a non-participant observer. Instead of observing behaviour, one is observing discourses: and the similarity does not end there. His belief that it is possible to observe society by 'standing on the edge of it' closely resembles the Positivist belief that it is possible (for the researcher) to make detached observations of a research subject (Taylor, 2007). It may not be possible for a person to stand outside (and objectively observe) society, but they can (objectively) observe an impersonal construct such as a discourse. This is as close as Foucault dares come to Positivism and the philosophical debates about knowledge (and whether it can exist independent of the knower). It is rather ironic that a methodology from a rebel of the establishment, such as Foucault, should (in a way) resemble the most conservative of all methodologies – Positivism.

Logical Positivism is the dominant methodology in both the social sciences, and natural sciences. It is what most people think of when they think of a 'scientific approach'. For those of you who have seen *'Star Trek'*, Vulcans are exemplary Logical Positivists: they remain emotionally detached from any problem, consider all the relevant (but only the relevant) variables, weigh the evidence objectively. This (process) 'naturally' leads them to the most probable conclusion, which they report impartially and without emotion. They are simply servants of the process. This is very much how Logical Positivism sees scientific research. (I am, of course, aware that we are talking about a people from a different race here!) The fact that this illustration is fictional does not make it any less valid (as most philosopher will tell you). Vulcans *are* exemplary Logical Positivists, and the fact that they are fictional only serves to make the criticisms of Logical Positivism all the more pointed. Many Feminists have asserted that it is almost impossible for a researcher to remain emotionally detached from the situation, and it is questionable whether people are capable of weighing evidence objectively.

Foucault has a very original style of writing, and (where possible) it is certainly worth reading Foucault in the original, just to get an idea of how he thinks. The following is a quote from his book *'Madness and Civilization'*. It shows how he has applied (t)his idea of a Discourse to the care and treatment of mentally disordered individuals. For Foucault, the Discourse is a form of covert control. (Those who know how to use the discourse have power over others.) Here, he is making the point that the Discourse (the real form of control) is virtually untouched by the 'civilising' of the treatment of mentally disordered individuals:

> What disappeared, in the course of the eighteenth century, was not the inhuman rigor with which madmen were treated, but the evident necessity of confinement, the total unity in which they were situated without difficulty, and those countless threads that wove them into the continuous texture of unreason. Madness was set free long before Pinel, not from the material constraints which kept it in the dungeon, but from a much more binding, perhaps more decisive servitude which kept it under the domination of unreason's obscure power. Even before

> the Revolution, madness was free: free for a perception which individualized it, free for the recognition of its unique features and for all the operations that would finally give it its status as an object…
> This was the first stage: to reduce as much as possible the practice of confinement with regard to moral transgressions, family conflicts, the most benign aspects of libertinage, yet to leave it untouched in its principle, and with one of its major meanings intact: the internment of the mad. This is the moment when madness actually takes possession of confinement, while confinement itself is divested of its other forms of utility.
>
> ***Foucault, as translated by Richard Howard, 1965:234-5***

Therefore, according to Foucault, this book is simply articulating the Discourse of social research – data becomes defined by the descriptors used in social research because it is being used in the context (Discourse) of social research, and that is how social researchers think of it. Even his own idea of a Discourse is itself a Discourse, in the Foucauldian sense of the word. And the fact that his own theory can even explain its own existence makes it all the more persuasive.

The main problem with Foucault's methodology, as I see it, is that it leaves little scope for expansion; all a researcher can do is to expand on the areas to which this basic set of principles can be applied. From a researcher's point of view, it has negative heuristic value; the concept of the (Foucauldian) Discourse explains everything, thus doing away with the need for any further research. In this respect, it does not contribute *that much* to 'knowledge'. The criticism that Foucault raises about Discourses in general, namely the constraining and self-defining nature of them, is also a criticism of the Discourse of Foucauldian Discourses. And I think Foucault himself recognised this, and his technique of 'standing on the edge of society' may have represented an attempt to escape the Discourse he had created.

Foucault has been the only philosopher, in recent years, to have attempt a 'grand theory' of human behaviour, and, for that, he is to be commended. Academics seem to be either passionately for, or passionately against Foucault's ideas. It is simply not possible to only be *partly* committed to Foucault's philosophy; it constitutes a complete way of conceptualising human experience, which is so radically different from anything else that it is simply not possible to relate it to any other way of seeing human experience. In this respect, his work is unique. For those who believe in (and follow) his ideas, Foucault is/was a genius, and a rebel, who gave the academic establishment something to think about.

In many respects, the uncomfortable and constraining nature of Foucault's Discourse alerts us to the fact that the terminology, identified above, is constraining (and this is as true for Foucault's own philosophy as it is for the Discourses that are the subject of his writings). Foucault would argue that you can not use it without it controlling and defining you; any attempt to rebel simply contributes to the Discourse. Discourse analysis, as it is used by researchers, often involves deconstructing what Foucault refers to as a Discourse in order to understand how it came to be constructed. By making visible the way in which it came to be constructed it is possible to understand the construction process and thus control it by making it the subject of the analysis. By deconstructing the way in which research data is defined, it has made the discourse visible and thus less powerful. It is the data and what it tells us that is important; any structure should only be a framework that helps in this process.

The Context of Social Research

In recent years, there has been a *slight* shift in emphasis away from the rules governing the quality and nature of data towards the importance of the data itself. People are more inclined to ask *'What does this tell us?'* rather than *'Is it good data?'* (i.e. valid and reliable). Issues of validity and reliability are still important but we now have more of a *'half full'* rather than a *'half empty'* attitude towards data. In the past, researchers would ask themselves 'is the data valid and reliable?', and if the answer came back 'no', then the data was largely dismissed as having little or no value. Today, there is more of a tendency to look at how data *can* be used (with the caveats that it may have been produced with a certain agenda, for a certain reason, or have been edited in a certain way), rather than to dismiss it for its failure to meet certain parameters.

There are a good many (psychology) research methods texts which spend a great deal of time (as I do) discussing ways in which researchers can avoid the situation where they, consciously or unconsciously communicate, to the research participant, what they are expecting to see (by way of responses from them) and this is often reciprocated by the research participant themselves. One of the most common expressions one hears, as a researcher, (from research participants) is *'is that what you wanted?'* And there is a sense (on the part of the research participant) that if they (the research participant) do not give the response that the researcher is expecting then they are somehow letting the researcher down.

With the advent of reality television and social networking sites (Facebook, MySpace, Bebo and YouTube), people have become more attuned to the idea of impression management, and the idea that things may not always be what they appear. Both lay (there has been an explosion in people researching their family trees in recent years) and professional researchers, alike, seem to have a (sixth) sense that data which is incidental (rather than intentional) is more trustworthy. Anyone who has watched documentaries on archaeology, genealogy, or investigations into the history of houses and/or shipwrecks will have noted the use of expressions such as 'giving up its secrets' (with reference to the object under investigation). It is almost as though we (now) think of every object, organisation or person as having a side to them that is public and a side which remains hidden (secret), and we are challenged to discover the 'private' side. (How many tabloid articles or TV programmes have the title or 'hook' 'the private life of…?)

Any researcher who has lived through at least some of the schools and movements which have influenced a particular discipline (see Taylor, 2007), will tell you that research has trends, and this recent innovation in new technology shows that the way in which people conceive of the research process, *itself*, is subject to trends and influences in the wider society. It is not simply how one goes about conducting research (of which the Feminist methodology of the 1980s is a good example) but what one believes one in doing in the process of conducting research. We now imbue inanimate objects with the same rational and motivation to present and hide certain information, from us, as our 'friends' do on social networking sites. This way of conceptualising the research process would have been unthought-of before the advent of such technology.

This said, it is not, in itself, an invalid activity to look at how such 'personas' are constructed, and many researchers have done just that. The most famous of which is Erving Goffman whose work on the subject includes titles such as: *The Presentation of Self in Everyday Life,* 1956*; Encounters,* 1961*; Behaviour in Public Places,* 1963*; Stigma,* 1963*; Interaction Rituals,* 1967*; Asylums,* 1974*; Frame Analysis,* 1975*; Gender Advertisements,* 1979 *and Forms of Talk,* 1981 (Preface to *Stigma*, 1986 edition). Nor is it 'wrong' to seek out data that does not have this element of 'impression management' to it. In fact, it has become a 'modern take' on the research process to look at how social phenomena have been socially constructed, and this preoccupation with 'discovering secrets' even applies to research sources themselves! Modern researchers are not content to simply accept the data that their forerunners have collected for them; they want to know about the people who collected this information, and their 'secrets'. This can be seen in the following two examples.

The Case of Refacement Graffiti

Refacement graffiti (Posener, 1982) has recently re-emerged as a form of social protest. For those of you who are unfamiliar with this phenomenon, re-facement graffiti involves the defacement of an advertising hoarding (billboard) in order change its meaning, usually to make a social comment by deconstructing (or reconstructing) socially constructed images. The defacement may take a number of forms [pages refer to Posener, 1982]: words added by graffiti (p36 top), tearing out parts of the advert (p25), or pasting words and/or images onto the original, usually in a sympathetic typeface (p24 bottom). (Or any combination of the above). This refacement is not simply intended to change the meaning of the advertisement; it is intended to make a social comment about the product (p91 top), the portrayal of a particular social group in advertising that product (p24 bottom), or the status of a particular social group in general (p36 top).

When refacement graffiti first emerged (in the early 1980s), it was amid a climate of anger and rebellion: Feminism was gaining strength, there were riots in Brixton and Toxteth, the best selling record of the decade (*'Relax'* by *Frankie Goes to Hollywood*) was one which had been banned by the establishment, and refacement graffiti was part of this protest culture. The examples of refacement graffiti, detailed in Jill Posener's book *'Spray It Loud'*, are manifestations of the anger that people felt at the time. Arguably, the refacement graffiti is characterise more by anger than wit, although one can not help but admire the ingenuity which went into some of these 'refacements' (p36b). The main themes were, clearly, female subordination (in both the selling of products and society generally), and the dangers of smoking. In addition to the examples of graffiti themselves, Posener provides quotes to show how people feel (felt at the time) about these issues (p13):

> This ad was opposite my place of work. I had to stare at it out of the window. A colleague and I went out and added the graffiti. You can see there are two handwritings! It was a way of taking over the poster. You have to have a lot of money to afford billboards like that. We wanted to reclaim the open spaces that have been colonised by advertisers. By writing angry but humorous graffiti, we were also making a point that ad agencies don't have the monopoly on wit. It feels great to see it reproduced everywhere. It's made the point that women can do something instead of just seethe.
>
> ***Reported in Posener, 1982:12***

> We want to hit men where it hurts – their pockets. They make money out of our bodies and lives. Every time I hear about another woman murdered or raped I feel I could commit violence. But I am scared and I don't like violence. And, in the end, people just misunderstand our actions unless we're very careful to publicise them....I want to see men change. I want them not to be violent...
>
> ***'An Angry Woman' in Posener, 1982:30***

> Our targets so far have only been the tobacco industry's ads. Cigarettes are the cause of most preventable illnesses in the UK. Most people get moralistic about kids starting smoking, and yet we're surrounded by ads which suggest it's cool and groovy to smoke. Hoardings dominate our environment. And they work on the unconscious, persuading people to buy without them quite realising it. COUGH UP does n't have any money. We can't buy hoardings. So we have to take them. We'd like to see an end to all tobacco advertising....It's all about "image". We'd like to see everybody who objects to advertising taking the same action as we do. In [s]tead of just passively accepting things, we can get up and do something.
>
> ***Reported in Posener, 1982:90***

Twenty years on, the methods are still the same, but the 'causes' (for which the protesters are fighting) and 'messages' are rather less well thought out. Recent examples of refacement graffiti show an advert for *'Coca-Cola'* in which the words *'Life tastes good'* have been amended to *'Life tastes good without'*, and an advert for *'The Mirror'* in which the words *'Tabloid accused of putting news first'* have been changed to *'Tabloid accused of putting news last'* (*The Guardian Education Section*, 5 June 2001:16-17). Here the 'cause' is the anger of students at: '(a)dvertisers (who) try to play to the cynicism of students by being funny...Students are tired of the image of students having sex and drinking, even if it is true' Sharon Williamson of Deadline (Advertising) (The Guardian Education Section, 5 June 2001:16-17):

> We object to so much advertising being thrown at us. At least with billboards it can be a two-way conversation. Here we wanted to say that a happy life isn't about soft drinks.
>
> ***'Tom', an engineering undergraduate from Warwick University, reported in The Guardian Education Section, 5 June 2001:16***

Today's cause is 'jamming corporate culture' (*The Guardian's* interpretation of Kalle Lason, founder of the Canadian branch of this movement), and highlighting the fact that such billboards are an eyesore:

> It's going to be changed in a week so there is no permanent damage. Plus: who is defacing whose property? When were local residents asked "Can we please have

a big billboard next to your house?" It's local councils and advertising making money out of us.

'Tom', reported in The Guardian Education Section, 5 June 2001:16)

Not only do these re/defacements lack the wit and political 'punch' of the earlier protests, but the refaced messages are actually *less* inventive than those of the original adverts. The re/defacement of the 'Coca-Cola' advert only serves to highlight how the use of iconography has made words virtually redundant in contemporary advertising. This graffiti only serves to highlight how the 'refacer' has missed the point of contemporary advertising (see, for example: Dyer, 1992, *'Advertising as Communication'*; Leiss, Kline and Jhally, 1990, *'Social Communication in Advertising: Persons, Products and Images of Well-Being, Second Edition'*; and Williamson, 1993, *'Decoding Advertisements: Ideology and Meaning in Advertising'*). Moreover, the irony of The Mirror's point about news in tabloid newspapers (also mimicking the 'read all about it' style of newspaper sellers), is turned into a rather literal, but obvious, statement about the same.

In response to the graffiti, Andrew Coker, the director of communications for Coca-Cola Ltd UK, is reported to have said: "It's a shame these students don't use their obvious creative talents in some more creative and constructive way." (The Guardian Education Section, 5 June 2001:16). Arguably, the executive from Coca-Cola® is being rather generous in describing these individuals as people with 'obvious creative talents'. The biggest disappointment here is not the way in which these students have vandalised other people's property, but that they can not do better than this, either by way of a 'cause' (which seems to be very vague and ill defined), or their ability to 'out-think' the advertisers. See what you think:

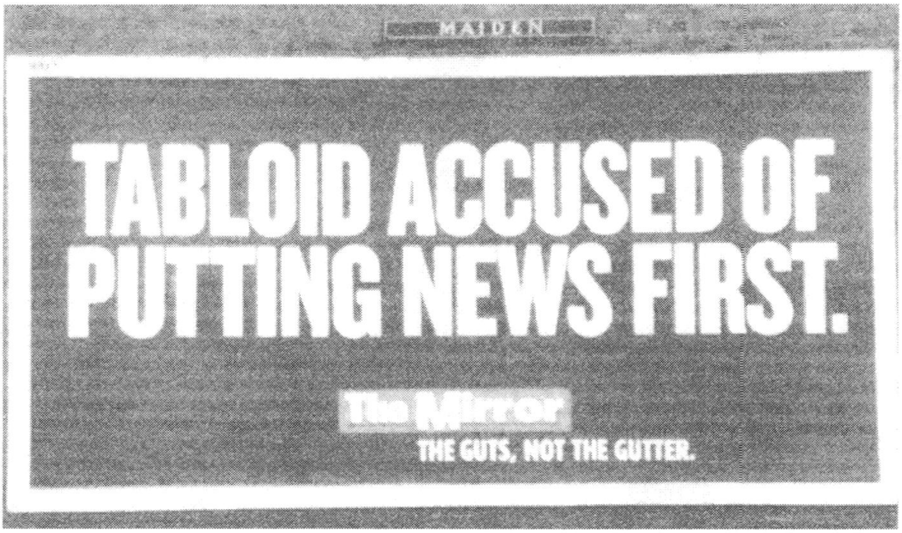

Acknowledgements to Joe Plomin and Guardian Newspapers Ltd

Acknowledgements to Joe Plomin and Guardian Newspapers Ltd

You might like to know that the permission to reproduce these images came in record time (six days), but this probably has little to do with the fact that the images, which are published in one newspaper *(The Guardian)*, make a mockery of one of its rivals *(The Mirror)*. I do have to say that, in my experience, students work very hard for their degrees, and I am appalled at the suggestion that they spend much of their time drinking and having sex (and please buy my book).

The *Archives de la Planète (ADLP)* and the *Eugenics Record Office (ERO)*: A Tale of Two Archives:

The *Archives de la Planète (ADLP)* was created by Albert Kahn between 1912 and 1931. Kahn was an architect who lived and worked around the turn of the (20th) century, and made a fortune as an architect. It was, perhaps, this work which alerted him to the fact that the world was changing; it was becoming more industrialised, more homogenised, more 'developed'; developments we now refer to as modernity. He somehow sensed that this would engulf and destroy the individuality of different nations, cultures and peoples, even lifestyles, and he set about creating a record of those cultures and lifestyles which he feared would be lost as a result of this social and industrial progress; in his words:

> '(T)o put into effect a sort of photographic inventory of the surface of the globe as inhabited and developed by Man at the beginning of the twentieth century.'
>
> ***Kahn quoted by Castro, 2006: Note 8, Jumpcut website***

His *Archives de la Planète (ADLP)*, which are now housed in the *Musée Albert Kahn*, contains some 72,000 colour images, and over 100 hours of cinematic footage of the peoples of 48 nations (ibid).

It is ironic that, for a man who has given us so many images of different nations and cultures, only one or two images of Kahn, himself, exist; a point which is not lost on contemporary historians (BBC *'The Edwardians in Colour: The Wonderful World of Albert Kahn'*, broadcast November 16th 2007, see also Castro website and link). What is interesting about this is that the way in which we seek to understand Kahn, and the way in which he sought to understand others, throws into sharp relief the differences in the approach to research that has taken place over the last 100 years.

Arguably, the archive itself tells us something about Kahn. Kahn was the one of the first people to employ *professional* photographers, and when one considers that photography, not to mention colour or cinematic photography, was in its infancy (work on the archive started around 1912), then to find and/or train a specialist in this field would have taken considerable resources. Moreover, not only did Kahn pay for his photographers' training and wages; he also paid their travelling expenses (from France) to places as far a field as the Far East and Africa (at a time when steam trains and ocean-going liners were the only transport available). Similarly, he furnished his photographers with what was (then) state-of-the-art equipment – autochrome cameras; the world's first user-friendly colour photography camera (invented only a few years' previously by the Lumière brothers). If we translate this into what state-of-the-art technology costs today then, all-in-all, this would have been a massively expensive undertaking.

The fact that he appeared not to seek any publicity for the project, or to take any credit for what he was doing, suggests that he derived no personal benefit from the archive, nor did he appear to have any (personal) use for it. One could argue that he may have used it as a sources of inspiration, but the images are of people rather than buildings (you will recall that Kahn was an architect). The evidence suggests that Kahn was well-connected and independently wealthy, so he did not need to exploit it as a source of income, and he would have almost certainly made a loss on the project anyway. One gets the feeling that Kahn felt he was engaged in a race against time – he had perhaps seen how people were willing to sacrifice their traditional lifestyles for the improvements brought by industrialisation and mass production (and who can blame them?), and so set about preserving, at least in photographic form, those cultures and lifestyles which he feared would be lost. Not as exhibits in a zoo but as a service to those nations and peoples.

We can ask ourselves 'How accurate a record (of these peoples) is this?', 'How representative are the people portrayed in these images?', or even 'How can we analyse this?' but to do so would miss the point of what Albert Kahn was trying to achieve. Arguably, the example that Kahn, himself, sets is one of the most valuable things to be found in his archive (but this is a modern researcher's perspective on the situation). Unlike that of the *The Eugenics Record Office (ERO)* at Cold Spring Harbour (also created around this time), there is (apparently) no hidden agenda to the archive – its point is precisely that stated by Kahn (letter above).

The ERO archive was created out of a fear of the 'criminal/working classes' rising up and taking over the country (the United States of America). At the time, Marxism predicted that the industrialised, capitalist economies would be the first to fall to communism. There was a fear, among the moneyed classes, that they would lose their wealth and position should such a revolution occur; a fear which helped the architects of the ERO to secure funding for the project detailing family 'pedigrees' (Allen, 1986). We will look at the ERO and its objectives in more detail, later, but, for now, suffice to say that eugenics is the science of breeding people, and the work of the ERO did not make a *positive* contribution to this endeavour. The ERO archive was closed down shortly after the start of the Second World War (and this is no coincidence), because the data was so flawed that it was considered to be of little or no scientific value (Allen, 1986).

Both of these archives survive today but they do so as much as a record of the kind of society and mindset that created them (and the lessons that can be learnt from it), as they do for the material they contain. It is nothing new to suggest that research is influenced by its political and social context but, perhaps, more than anything else, these two archives show how the motivation behind the creation of that knowledge can have an effect on the data collected (and the ultimate value of that data). As much as Positivism may attempt to deny the role of the researcher; and as much as Foucault (see above) may attempt to get around the idea by claiming to 'stand on the edge of society', the researcher is an integral part of any research project. This does of course re-ignite philosophical debates into the nature knowledge, specifically the ***Realism v. Phenomenalism*** and ***Epistemology v. Ontology*** issues:

Realism v. Phenomenalism – is there a reality independent of our ideas and experiences (that we can come to know) or must 'knowledge' always be based on our understanding of phenomena as we experience them? (after Hammersley, 1989:6)

Epistemology v. Ontology – Should we found our investigation on assumptions about how knowledge is possible or the nature of the world that we seek to understand? (after Hammersley, 1989:6)

In the 1980s, the Feminist movement argued that it was wrong to write the researcher's thought and feelings out of the research process (largely as a reaction to the apersonal, Positivist methodology used in many of the sciences, particularly psychology). These examples show how even the most self-aware researcher will, eventually, be betrayed by their motives. As resources such as this are made to 'give up their secrets', the person behind the research will become as much a subject of research as the work itself. The motivation of the researcher is a factor in the research process, it can also be one of its biggest flaws (together with the fallibility of the research participants themselves).

Flaws in the Research

A piece of research can be 'flawed' in many ways. These 'flaws' include everything from the deliberate falsification of data, to the characteristics of people who volunteer for research. Cases of academics deliberately falsifying their results are rare, but they do happen. It is more likely that the researcher has been careless or naïve with regard to some aspect of their (research) design or procedure. This does not mean that the 'flaws' are always obvious or visible. The research process is, very often, an interpersonal situation, and, in the course of conducting a piece of research, the researcher's non-verbal behaviour might clue a participant in to what they are looking for. A sigh of relief may be all it takes to cue a participant in to what the researcher is looking for. The participants may also be, actively, trying to work out what the researcher is looking for. This may be out of a desire to see the research 'succeed' (or fail), for the researcher, or to not disappoint/let the researcher down, by failing to 'perform' as expected. There are many such sources of error, including; 'the experimenter's expectations and personality..., subjects' personality..., their awareness of the experimenter's intent..., or their concern that they are being evaluated' (Boring, 1969:6).

It is also the case that a researcher's expectations may lead them to see what the want to see. An example of a researcher seeing what they wanted to see (or more precisely removing the data that they did not want everyone else to see), happened recently in America. Dr. Dean Hamer claimed to have found evidence for a 'gay-gene' (passed on naturally from mother to son). The discovery was celebrated by gay rights groups, across the world. Some even went as far as having T-shirts, with the words *'Thanks for the Genes Mom'*, printed on them. Overnight, Dr Hamer became a celebrity, a potential award winning scientist, and a best selling author. The discovery was considered so significant that his paper was accepted for publication by the academic journal *'Science'* (an extremely prestigious scientific publication). As you may already have guessed, Dean Hamer was, himself, gay, and it was not long before the academic community began to wonder whether his own sexual orientation may have influenced his research. Shortly afterwards, it was discovered that he had 'selectively report(ed)...his data in ways that suited his own hidden agenda' (Almond, *Mail on Sunday*, 9 July 1995).

A similar, but less deliberate, example of a researcher seeing what the wanted to see, can be found in the work of William Sheldon. Sheldon worked around the first quarter of the twentieth century, and he was interested in investigating whether or not there was an association between body type and temperament. His method involved taking photographs of 4000 male undergraduates, classifying them according to three body types: *Endomorphic* (round and soft), *Ectomorphic* (narrow and skinny), and *Mesomorphic* (strong and muscular). Sheldon then took a list of 650 personality traits, narrowed them down to fifty, and then reduced them further to three principle categories (by a statistical procedure known as Factor Analysis, which is discussed in *Interacting with Statistics*, Taylor 2007): *Viscerotonic* (loves comfort, needs approval), *Cerebrotonic* (tense and hesitant, a loner, not fond of action), and *Somatotonic* (noisy and aggressive, always on the move). He found that an Endomorphic body type correlated highly (again the technical meaning of this is explained in the above text) with a *Viscerotonic* temperament, a Mesomorphic body type correlated highly with a *Somatotonic* temperament, and an *Ectomorphic Mesomorphic* body type correlated highly with a *Cerebrotonic* temperament (see Pitts, 1987:543-4).

Unfortunately, when it came to replicating these findings, no one could match Sheldon's original correlations (of around +.80). (For the purposes of this illustration, it will suffice for you to know that a correlation is an index of the strength of the relationship, ranging from 0 to 1: 1 being the highest.) Sheldon scored participants both on body type and temperament, whereas subsequent researchers used separate people to score for body type and temperament. This suggests that Sheldon may have been seeing what he wanted to see (Pitts, 1987). (These are known as **'experimenter effects'.**) Sheldon was not trying to deceive anyone; it is just that his expectations exerted a stronger influence than he realised. It is tempting to believe that, because Sheldon was working in the early part of the twentieth century, he would have been more naïve than us. He may have been more ignorant of what we call **'demand characteristics'**, but there is no reason to think that this made him any more vulnerable to his own expectations than researchers are today. In order to reduce the risk of an observer seeing what they want to see, they should (ideally) be unaware of **(blind to)** what the researcher expects to find.

By the same token, participants are sensitive to what Orne refers to as 'demand characteristics'. These are cues, which clue the participant in to what the researcher is looking for (and there is no reason to believe they are peculiar to experimental research). In order to eliminate demand characteristics, the researcher should (ideally) employ assistants who are naïve (or 'blind') to the aims of the study. Even then, this does not stop the participants looking for clues as to what the research is about. Even if the researcher is able to control for demand characteristics, within the research setting, participants will still have certain expectations about the setting and the nature of 'scientific' research. This is worth bearing in mind, particularly if the research involves observing social interactions: as Orne points out

> Their (participants) understanding of the situation is based upon a great deal of knowledge about the kind of realities under which scientific research is conducted, its aims and purposes, and, in some vague way, the kind of findings which might emerge from their participation and their responses.
>
> ***Orne, 1969:144***

Even when a researcher is confident that they have controlled for 'experimenter effects' and 'demand characteristics', there are other factors which may influence the results. These include things such as the characteristics of the sample chosen. For example, those who volunteer to participate in academic research, may differ from those who do not. In the 1960s, two researchers, Rosenthal and Rosnow, reviewed the literature on this subject: their findings can be found in the table overleaf.

Variable	*Findings*
Sex	"Eight of the studies discovered that females volunteered more than males, while the remaining four studies found the inverse." (p.68-9)
Sociability	"(I)t would appear that volunteers, especially females, are higher in sociability than nonvolunteers." (p.73)
Conventionality	"These results are not unequivocal, but in general it would appear that volunteers for behavioral research tend to be more unconventional than nonvolunteers. Six studies support this conclusion. (However, two others find no difference, and two others report the opposite relationship.)" (p.79)
Anxiety	"On the basis of these seven studies one could certainly conclude that, at least in terms of manifest anxiety, volunteers are usually no different than nonvolunteers." (p.81)
Psychopathology	"Thus far the results of studies of volunteers for medical research agree rather well with one another, consistently tending to show greater psychopathology among volunteers than nonvolunteers." (p.87)
Intelligence	"For all four studies combined, volunteers were still found to score higher on intelligence than nonvolunteers." (p.87) "When the definition of 'intelligence' is in terms of school grades the relationship of volunteering to intelligence becomes more equivocal." (p.89)
Social Class	"There is a high degree of correlation between amount of education and social class, the latter defined by occupational status. One would expect, therefore, that in survey research those subjects having higher occupational status roles would be more likely to answer questions and to answer them sooner than subjects lower in occupational status....At least for the range of occupational statuses noted here, higher status role occupants are more likely to participate in the survey research process than those lower in status." (p.91)
Age	"(O)ften there is no significant difference between the ages of volunteers and nonvolunteers. However, when differences are found they most often suggest that younger rather than older subjects are those who volunteer for behavioral research." (p.92)

After Rosenthal and Rosnow, 1969:68-92)

Clearly, the characteristics of the people who volunteer to take part in research, is something over which a researcher has no control. Unlike demand characteristics, it is simply something that the researcher has to be aware of. Academics spend a great deal of time looking for (and seeking to eliminate) alternative explanations (for their findings), and flaws in the research are just part of this. This is part of what we call research design.

The Implications of Research Design

The research design (method or source chosen) can be thought of as the way in which the researcher goes about looking for the answer to the research question, and this is a product of the way in which the researcher 'views' the research question. Moreover, the way in which the researcher views the research question is a reflection of those (philosophical/methodological) influences (discussed in *'Interacting with Statistics'*, Taylor, 2007). In other words, the research design is a reflection of the influence on the interpretation of the research question. While it may be argued that certain research questions (i.e. the question rather than the underlying constructs), 'naturally', require a certain type of research design, it is, perhaps, worth reminding ourselves that the influences, discussed in *'Interacting with Statistics'* (Taylor, 2007), include not only the theory, but the school/movement, the methodology, and the discipline one is using.

In many cases, the decision to adhere to a particular discipline, methodology, school etc. is one which the researcher may have made earlier in the life of the project, if not their career (and has subsequently forgotten that it was a choice). This is why it seems 'natural' to use documentary evidence when conducting *historical* research; why it seems 'natural' to use a *Feminist Methodology* when conducting research into vulnerable individuals; why it seems 'natural' to use an experimental *(Positivist)* design when conducting research into causal relationships. These only *appear* to be decisions pertaining to the research question, because the relevant philosophical/methodological decisions *have already been made*. In other words, some (philosophical/methodological) decisions have been made implicitly, making it appear as though the design is the 'natural' choice for the research. The research design (method or source chosen) does not, therefore, provide an answer to the research question; it is only a way of *looking for* that answer.

The design also determines the *nature* of information that will be used to answer the research question. Broadly speaking, the data can either be in the form of a discourse (a dialogue or text), or a numerical measurement (a scale or count). Elsewhere, texts on this subject (research methods) tend to describe research designs by the type of data they produce - 'qualitative' or 'quantitative'. By 'qualitative data', they mean a the kind of information produced by 'open' (unstructured) questions – so called because it is considered to have 'richness', thought to be lost when the responses are translated to numerical measures or counts. In practice, the distinction tends to be one of 'non-numerical'/'numerical' information. Designs, therefore, tend to be distinguished by the data they generate, and the analysis performed on that data, rather than the design *per se*. It is questionable whether there is anything to be gained from describing the designs in this way, or even if the 'qualitative/quantitative' distinction is practical. Irrespective of how the design/data/analysis is categorised, the task is still the same – to design a piece of research, which will (ultimately) enable one to answer the research question (as it has been interpreted). In this respect, the 'qualitative/quantitative' distinction can be rather distracting. The reason for making it here is to draw attention to the way in which the two types of data ('qualitative' and 'quantitative') are analysed. Even then, it can still be a distraction - the task should be thought of as the same (to find an answer to the research question), regardless of whether the data is 'qualitative' or 'quantitative'. The utility of this distinction is even questionable when it comes to the task of analysing the data. When it is reduced to its most basic elements, 'data analysis' is, essentially, the same (process) for both 'qualitative' and 'quantitative' data.

Data Analysis

'Data analysis' is a process of assessing the information and drawing conclusions from it (This is the same, regardless of whether the data is 'qualitative' or 'quantitative'.) At this point, it is important to make clear that 'data analysis' can, effectively, be broken down into two stages - 'determining what is *there'* and 'interpreting what it means' (in the context of the research question). (Again, this is the same, regardless of whether the data is 'qualitative' or 'quantitative'.) The *form* that (the stages of) this process (data analysis) takes, *does*, however, depend on the *nature* (qualitative or quantitative) of the data. If the data is 'qualitative', then the first stage of this process ('data analysis') is likely to be a form of categorisation. Regardless of whether this process (categorisation) is described as 'looking for emerging themes', 'a content analysis', or even 'a discourse analysis' it is, essentially, a process of categorising the data. These categories can be created either before, or after, the data is gathered. If this is done *after* the data is gathered, then it can either be done manually, or with a computer programme; some of the most widely used programmes are:

NUD*IST (Non-numerical Unstructured Data by Indexing, Searching and Theorizing) - which can 'auto-code' text as well as import table data.

Ethnograph - which is used in the management and analysis of text based data such as transcripts of interviews, focus groups, field notes, diaries, etc.

ATLAS/ti - which is used for text interpretation, text management and the identification of concepts in documents.

Code-A-Text - which was originally designed to help therapists identify thought processes and ideas in interview with clinical patients but has applications for analysis of field notes and questionnaires.

Neuendorf, 2001 companion website

The 'problem' with this (or indeed almost any) type of analysis is that the researcher is likely 'identify' those categories which their culture (if not their research perspective) has 'taught' them to 'identify'. One of the ways in which this stage of the process could be made more 'scientific' is through the researcher showing a greater awareness of their 'own' expectations; Feminists refer to it as **'reflexivity'**. (Reflexivity is a process of making visible the, normally unseen, beliefs and expectations of the researcher.) This may not make the analysis any more 'scientific', in a Positivist sense (of a 'value-free' science), but it does, at least, make it more 'scientific', in a Popperian sense of making the process visible, and amenable to criticism (see *'Conjectures and Refutations'*, Popper, 1976 edition).

Once the data has been categorised, it then has to be interpreted. The interpretation given to the *data* will, in part, be determined by how the researcher has interpreted the research question. This shows (one of the reasons) why it is important to be able to articulate the rationale of one's design. (This is why it is important to read it, as it has been written.) While this 'formula' may make 'interpreting the data' much easier, there is no 'formula' for 'interpreting the data', itself. An interpretation is just that – one person's 'reading' of the data. This does not mean to say that two, or more, people can not come up with the 'same' 'interpretation', or that one person's interpretation can not be 'better' than another's (given the data, of which it is an interpretation), merely that 'an interpretation' is, in part, an opinion. Clearly, the degree to which the data supports this interpretation is the 'evidence' that the researcher will use to support their explanation of the 'findings' (their answer to the research question). (This is discussed, in more detail, in my other book – Taylor, 2007.)

The difficulty in making 'qualitative data analysis' expectation-free, has led to the misconception that 'quantitative data analysis' (analysis based on statistics) is superior (because it is based on 'mathematics', rather than 'human judgement'). 'Quantitative (statistical) data analysis' involves describing the data, using standardised, statistical descriptors (for example, an average value), and then calculating the probability of the descriptor having that value, *by 'chance' alone*. If the probability of it occurring, *by 'chance' alone*, is very small, then this points to it being a note-worthy finding. While this may make it appear as though the process does not require the researcher to use their judgement, this is not actually the case. The researcher still has to decide which descriptor(s) to use (although this will, in part, be determined by the research question), and what is meant by 'a very small probability' (although this will, in part, be determined by the academic community). Clearly, these decisions *will* have an impact on the researcher's interpretation of the data, as well as their explanation of the 'findings'. 'Data analysis' is, therefore, the 'same' (in terms of its objectives and principles) *process*, regardless of whether it is being conducted on 'qualitative' data, or 'quantitative' data. With this in mind, let us now look at how we come by this data.

A Typology of Research Methods

When it comes to acquiring data in order to *answer* the research question, there is huge array of methods and sources available to researchers which enable them to do this. This array of methods can be, quite literally, baffling! A number of authors have attempted to organise this information in a way which makes it accessible to the reader. Many are logically organised, but this does not seem to facilitate the recall of the substantive information. Arguably, the reason why readers do not remember this information is because the structure is simply seen as a way of organising the chapters. In other words, the structure is being used to organise the substantive information, but it is not being used to aid the understanding of it. (For a discussion, look up the work of G. H. Bower, in any good cognitive psychology textbook.) The rationale for organising the substantive information in this way may be crystal clear, to an expert; but it may not be quite so obvious, to a novice. If this rationale is central to (or at least beneficial in) being able to understand the method itself, then those who can not follow the author's train of thought, are at a distinct disadvantage. For this reason, it is perhaps worth explaining the rational behind the typology used in this text (which it is hoped will be intelligible to those with little or no specialist knowledge in this field).

The typology that will be used here is based on the origin and medium of the data. These categories should be seen as a way of conceptualising research methods, rather than pigeon holes into which different methods can be placed. As a system of classification, it is very 'loose' (and some of the decisions beg an explanation), but as a way of organising the literature, it follows an intuitively logical route through the relevant literature. The typology used here classifies the various research method (primarily) according to the origin of the data: that generated by or on behalf of the **researcher**, the participant(s) **(personal)** information, or that in the **public** domain; and then according to the medium of the data: **written** or **verbal**. With some methods, there is a fine line between the information elicited by a researcher, and the information volunteered by a participant. Therefore, the categories: 'researcher', 'personal' and 'public' should be thought of more in terms of who can exercise the greater control over the information, rather than the origin *per se*. The use of 'verbal' and 'written', as discrete categories, does not preclude the possibility that some methods may elicit both types of data. Likewise, 'verbal' is not meant to preclude the possibility that some of the data may be non-verbal. (Methods, which elicit and/or record verbal data, usually elicit and/or record non-verbal data as well.) The 'verbal' category should, therefore, be thought of as 'non-written', rather than verbal *per se*. Before looking at the various methods, in detail, it perhaps worth looking at how they can be categorised, according to this system of classification. This is displayed in the table below. (Those of you who are already familiar with these methods, can perhaps begin to see the logic of the structuring.) One chapter will be devoted to each of the main categories: 'researcher', 'researcher elicited personal', 'personal', and 'published'.

Researcher Written	**Experiments, Field Experiments and Quasi-Experiments**
Researcher Verbal	**Non-Participant Observations, Participant Observations and Ethnographies**

Researcher Elicited Personal Written	**Surveys, Researcher Elicited Correspondence & Diaries, Diary Interviews**
Researcher Elicited Personal Verbal	**Interviews and Focus Groups**

Personal Written	**Personal Records and Documents, Photos, CVs, Diaries, and (Auto)Biographies**
Personal Verbal	**Oral Histories**

Holistic	**Case Studies**
Published/Public Written	**Archives and Publications**
Published/Public Verbal	**The Media**

2: Research Methods and Sources – Researcher Initiated

Introduction

As mentioned, this typology has *not* been designed for the purposes of identifying the characteristics of a particular method (these will be discussed in the chapters themselves), but to show how different methods have evolved in response to different types of research questions. The experiment effectively evolves into a field experiment, which evolves into a quasi-experiment, which evolves into a non-participant observation. Just as, in the animal kingdom, there are families of animals, there seem to be 'families' of methods, which have certain methodological similarities. For example, participant observation and ethnography are concerned with 'lived experiences'. One can almost deduce what kind of research is being conducted from the (methodological) questions being asked. Obviously, this is something which could be discussed at length, but the purpose of discussing it here is simply to elaborate on the rationale of the typology. To this end, let us now look at the first of these methods – the experiment.

Researcher Written	**Experiments, Field Experiments and Quasi-Experiments**
Researcher Verbal	**Non-Participant Observations, Participant Observations and Ethnographies**

Researcher - Written

Experiments

The experimental method is classed as 'researcher written', because the researcher is the person with most control over the data (which they then recorded, in some form or another). While the image of an experimenter recording responses, with a pencil and clip-board, may be somewhat dated, this is still the history of the method. The recording devices, used now, may be more sophisticated, but the data still tends to be a response, written/recorded on a permanent medium. This can be contrasted to data which is the result of the researcher's interaction with others (as would be the case with a piece of ethnographic research).

As a researcher, it is important to understand the difference between what the term 'experiment' means to a researcher, and what the term 'experiment' means in popular culture. This is important because the term 'experiment' has a number of negative connotations which may cause prospective participants to become apprehensive about participating in the research. A researcher has to understand not only what these misconceptions are, and how they may have come about, but how to address them in participants who may not even know that they have these misconceptions. Conducting an experiment is, therefore, as much a matter of being able to reassure participants, as it is a matter of collecting the data. To a researcher, 'an experiment' is simply the name given to a particular method of gathering (generating) data, but, to a participant, it is part of the mysterious world of 'the scientist'. Investigating these misconceptions is a useful way of investigating the method itself. By addressing these misconceptions, it becomes possible to identify the parameters of the method, and, from this, it becomes possible to show how the method is used in social research.

The Misconceptions

One popular misconception is that they (the participant) have not been chosen at random, or have been chosen to participate for a particular reason. People are asked to participate in experiments, not because the researcher is interested in *them* personally, but because they are assumed to be 'representative' of their social group, or likely to produce a 'typical' response. (To enable the findings, of the research, to be applied **(generalised)** to that group of individuals.) An experiment is not about the responses of *individuals*; it is about the *aggregate* of individual responses. Participants should not, therefore, take the experience personally. As far as the experiment(er) is concerned, the participant is simply there as a 'representative/typical individual'. Even when a researcher is interested in individuals who have certain attributes, or are members of a particular social group, the participants are still only there as representatives of that group, not as individuals in their own right. This is not, however, without its ethical implications. Feminists, in particular, would find this approach (to research) objectionable, because it dehumanises people. It does dehumanise people, but, by the same token, there is no reason for participants to think that the research is about them personally. While this may violate the participant's right, to be seen as an individual, it respects their right to privacy.

This point also touches on another popular misconception – the belief that people (participants) are being experimented *on*. The expression 'researched on' implies that the participant is being *used* as a foil for the experiment, which implies that the participant is powerless. It perhaps goes without saying that this is something which the researcher needs to address – not only is it a barrier to ethical research; it is a breeding ground for demand characteristics. There are many issues here, in particular, the question of how much participants should be told. When it comes to 'informing participants', researchers tend to think of things like: the aims and objectives of the research, or what is required for 'informed consent', rarely do they entertain the idea that it could apply to the method itself. Part of this will involve informing participants of what their contribution, to the research, will be. Not only does this show that the researcher is willing to treat them as an equal, by sharing some 'technical' knowledge with them, but it enables them to see why they have not been told about other aspects of the investigation. It is important to get away from this idea that the participant is a 'guinea pig'.

In this respect, there is one connotation (of 'an experiment') that can be quite damaging – the idea that 'an experiment' is something which has not been tried before. The inference that it has not been tried before, suggests that the researcher does not know what is going to happen. In reality, this research question, which may have never been asked before, may not be all that much different to one which has been asked before. A great deal of research is designed to investigate the conditions under which a theory applies. Clearly, if participants were told that the research is simply to see whether the results of a study, conducted abroad, can be applied to British participants, they may not be as enthusiastic about it, as they would, if they were just left to imagine that it is this piece of really innovative research. (The problem comes, of course, when they start to imagine other things.) The implication that it has not been tried before, also suggests an element of speculation. 'An experiment' is often used as a euphemism for solving a problem by 'trial and error'. (Experimenting with different solutions.) This process of 'trial and error' is certainly *not* something which *academics* would associate with 'the experiment' (as a research method.) Any variations in (experimental) design are made systematically, and with a good justification for doing so. Experimentation is about the prediction of causal relationships, not guess-work. This brings us to another connotation of the word 'experiment' – that of a systematic procedure. The stereotypical image of 'an experimenter' is of someone working in a laboratory, methodically varying the amounts of different chemicals in order to arrive at the correct formula. Although this image may make one think of certain classic horror movies, academics would certainly not want people to no longer associate an experiment with a methodical procedure; and this brings us on to the nature of the experimental method itself.

The Method Itself

The experiment is a method designed to investigate causal relationships. The defining characteristic of a causal relationship is the ability to predict a particular outcome from a given set of circumstances. Saying that an event is caused by another, is the same as saying that the second event would not occur without the first. To put this in the context of a research question, a particular theory or hypothesis (research question) may predict that altering one specific aspect of the situation **(a variable)**, the outcome (usually another variable) will be altered in a predictable fashion. In order to investigate such (causal) relationships, one needs to be able control the research situation to the extent that this specific relationship can be 'tested'. (This is what the experimental method allows one to do.) This procedure suggests that certain variables are distinct from others, and that they can be identified by their function. Given that 'variables' are an integral part of the experimental method (and that different variables have different roles), it is important to look at what they are.

The concept of a variable is simple, and yet difficult to explain. A variable is any descriptive which can take at least two sub-types/values. Gender is a variable, because one can be described as either male or female. Memory for a list of words is a variable, because it can be described/measured in terms of the number of items recalled. Reaction to a particular medical treatment is a variable, because it can be described/measured in terms of the degree of recovery. Variables are given different labels, depending on their part in the research. The **Independent Variable** is the variable, which it is believed will affect the outcome event. The **Dependent Variable** is the variable, which it is believed will be altered, as a consequence of altering the independent variable. In other words, the dependent variable measures the effect of **manipulating** the independent variable. These variables, may be expressed as predictions of causal relationships, such as: gender determines liking for sport, level of intelligence determines academic performance, liking for violent films determines criminal tendencies, age determines memory for one's route to school. In each case, the independent variable is hypothesised to have an effect on the dependent variable.

Not only must the variables be identified, but their meaning must be universally understood. For example, 'aggression' means different things to different people. Therefore, a study setting out to measure "the effects of gender on aggression", needs to say, specifically, what it means by 'aggression'. This (specific definition) is known as an **operational definition**. An operational definition is how a particular variable (or level of that variable) is being defined, for the purposes of that particular piece of research. For example, a child's ability to remember their route to school, could be measured in a number of ways: by asking them to show (take) someone the route, by asking them to trace the route on a map, or by asking them to draw a map of the route. In each case, the child's memory for their route to school, is being defined/measured in a slightly different way, and this could have implications for the way in which a particular dependent variable affects it. The measure may not be sensitive enough to distinguish between differences caused by the dependent variable. Similarly, it may rely to much on the child's competencies, in other areas, to produce a valid measure of their memory. Operational definitions are important because they make it possible for other researchers to **replicate** (reproduce) the research (and its findings), for themselves. (This shows that the findings are researcher independent.) If variables are inadequately defined, then any differences (between the two studies) may be due to how the independent (or the dependent) variable is being defined, rather than (say) cultural differences (and this would effectively constitute a **confounding** variable).

Researchers can not always guarantee that the research setting will be the same for each participant. Nor can they guarantee that the conditions will vary only in terms of (the differences in) the independent variable. A **confounding** variable is a variable, which differs along with, but is not related to, the independent variable. The presence of a confounding variable, makes it difficult to tell whether the change in the dependent variable is due to the difference in the independent variable, or the difference in the confounding variable. The researcher can not say, for certain, that (the manipulation of) the independent variable is having an effect on the dependent variable. Let us suppose that we are interested in investigating the difference in children's memories, for their route to school – specifically between children aged six and eight years old. If the group of six year old children contained a greater proportion of children who were taken to school by car, then this could have an effect on the results. It may mean that the eight year old children have more time to survey the route; it may also mean that the eight year old children have a kinaesthetic (muscular/physical) memory of the route, to go with their visual memory of the route. Travelling by car, versus travelling by foot, would, therefore, have a confounding effect (on age, the independent variable) on learning one's route to school. A researcher, therefore, has to be able to identify, and **control** for, confounding variables such as these.

Similarly, a researcher has to be able to control for demand characteristics and experimenter effects. It is difficult, for a researcher, not to give out subtle cues as to what kind of responses they may be looking for. A sigh when the participant gets it 'wrong', may be all that is needed to cue the participant in to what the researcher is looking for. A great many participants want to see the research 'succeed', for the researcher's sake. To this end, they will often try to 'decode' the research, in order to find out what it is that the researcher 'wants' (and be able to adapt their responses accordingly). The only way round this 'problem' is to disguise the (true) nature of the research, such that both the participants and the experimenter(s) are in ignorance of (**blind** to) the actual research question. Clearly, this raises some ethical issues. (Which most ethical guidelines suggest can be overcome by informing people, more fully, when the research is over.)

It can be difficult for a researcher to control for all these (**extraneous**) variables. In order to achieve this, some researchers have ended up producing some very artificial situations. A good example of this is the research into, what has become known as "The Cocktail Party Problem" (Colin Cherry, 1957:277-80). This is "the ability to listen to, and follow, one speaker in the presence of others" (Cherry, 1957:278). If one was trying to design an experiment, to investigate this phenomenon, it would be extremely difficult to ensure that the volume and content of the background (noise) is the same for each participant. Similarly, it would be extremely difficult to ensure that the volume and content of the 'attended to' conversation is the same for each participant. Cherry actually solved this problem by doing away with the party, altogether. Both the attended, and the unattended, 'conversations' were replaced by pieces of prose. This ensured that everyone would be at the same 'party', listening to the same 'conversation'. The pieces of prose were recorded, using a tape recorder, and the recordings were presented through headphones. This ensured that the presentation would be the same for everyone. Volunteers were asked to "separate" the two pieces of prose, by dictating extracts, as they were able to make sense of them. The participants were not allowed to write anything down. (For a review of his early experiments in this area see Cherry, 1957.) While this procedure may make for a highly controlled experiment, what is really left of the 'Cocktail Party'? Clearly, there comes a point where the research setting is so unlike 'real life', that one must question whether the results (of that research) can be applied to the real world. (Or, as researchers would say, it has little **ecological validity**.) This creates a trade-off between controlling for extraneous (possible confounding) variables, and retaining something of the research's relationship to the real world. One way of getting round this problem (of ecological validity), is to actually conduct the experiment in the real world. This is what is known as a field experiment.

Field Experiments

Many academics make the distinction between an experiment and a field experiment, as if they were two completely different approaches. Conceptually, the difference is not all that great. Experiments tend to be conducted in 'laboratories', where the research environment can be tightly controlled. These 'laboratories' vary greatly, in the degree to which they are 'equipped' for the purpose. A laboratory is a laboratory, not because it contains any specialist equipment, but because it has been designated as such. (It may be just an empty room.) *Field* experiments are experiments carried out in a non-laboratory environment. The Hawthorn study is a good example of a field experiment.

In 1927, Elton Mayo conducted a piece of research, in the Relay Assembly Test Room of the United States Western Electric Company, Hawthorn Works (Chicago). The effect, discovered in that experiment, is known, to this day, as the 'Hawthorn Effect' (see Blumberg, 1968 for a discussion). What Mayo actually set out to investigate was the effect, on productivity (in the manufacture of electrical relays), of changes in the working environment. He found that every change improved productivity (even reversals of earlier changes). Productivity improved, not because the physical environment leant itself to increased production, but because the management began taking an interest in the workers (through the conditions of the experiment). For the first time, the workers were being consulted about their working conditions, and the interest shown, in them, by the management and the company, was reciprocated through increased production.

This research could not have been carried out in a laboratory. It probably has not been carried out anywhere other than in that particular working environment. With the Hawthorn studies, the researcher was still manipulating one variable (working conditions) in order to see what effect it had on another (output). The only difference between this and a conventional experiment, is that the research is being conducted in a factory rather than a laboratory. Apart from this, the principles of the method are exactly the same.

The Quasi-Experimental Method

A quasi-experimental design is like an experiment, in that it is still investigating the effect of ('manipulating') one variable on another (in an assumed causal relationship), but the manipulation is not instigated by the researcher. It is largely academic, whether one sees it as a design in its own right, or as a variation on an experimental design. Graziano and Raulin's (dictionary) define 'quasi' as: "similar to" (1989:271); whereas the *Collins (Gem) English Dictionary* defines it as: "as if" or "as it were"; and the *Webster's (New World Compact) American English Dictionary* defines it as "seemingly". It is a small difference, but it is perhaps better to think of the research being conducted *as if* it were an experiment, rather than using a design that is *similar* to an experiment. A quasi-experimental design is used when a researcher wishes to investigate the impact/effect of a particular event or intervention. This intervention may be a medical treatment, a traffic calming measure, a new management policy, a community project, a course of counselling, a tax increase, an advertising campaign, a consciousness raising event, an anti-crime initiative, and so on. A variable is still being 'manipulated', and its effects (on a relevant dependent variable) are being investigated, but the researcher plays no part in that intervention.

The decision to implement an intervention, of the type described above, necessarily has to be made by someone in authority/with the authority to do so. That person is, therefore, accountable for the consequences of that decision. The fact that one of the design decisions has been taken out of the hands of the researcher, means that the researcher's ability to influence the design is as great as it would be in a true experiment. This may cause some researchers to argue that it is not the same as an experiment – the difference being the control that the researcher has over the design.

Academic debates not withstanding, the quasi-experimental method/design does have a place in academic research. Suppose that a new treatment for AIDS is going to be tested at the local hospital. The hospital has already decided who will receive the treatment, and who will receive a placebo. (A placebo has the appearance of the treatment, but without the active ingredient.) This will determine the benefits of the treatment, over and above the psychological benefits of receiving a *treatment*. However, the hospital is not prepared to disclose which patients have been assigned to which condition; either to the patients, or the nurses administering the treatment. (This is known as a double-blind procedure.) This has all the components of a good experimental design. All a researcher has to do, is to accommodate the (design) decisions which have already been made. As far as the design is concerned, it is irrelevant *who* manipulates the independent variable. The only concern for a researcher may be that they not know how the patients were selected for these trials, or the basis on which they were assigned to each condition. This aside, it is possible to proceed with the research, as if it were an experiment. In this case, we would expect to see an improvement in the health of the patients in the treatment group (from the pre-treatment to the post-treatment period), over and above that of any improvement in the health of the patients, in the placebo group (for the same period). This is a typical quasi-experimental procedure. There are variations on this design, but they are just variations (see Graziano and Raulin, 1989:271-281).

Researcher - Verbal

In this section, we move from methods, where the researcher *controls* the research setting; to methods, where the researcher places themselves *in* the research setting. In both cases, the researcher is able to exercise a high degree of control over the information gathered (this is why they are in the same category, according to this typology), but the written/verbal distinction is perhaps a little artificial. The non-participant approach to observational research is perhaps more akin to the approach of the experimental method, than it is the 'lived experience' approach of participant observation or ethnography. While researchers may distinguish between methods, on the basis of gross features of the design (for example: 'an experiment' or 'an observational study'), when one starts to look at the features of the various designs, in terms of the differences between them, some of these differences are not as great as they first appear. Classifying non-participant observation as a completely different type of method (to an experiment), is more likely to be an obstruction than it is to be an aid, when it comes to understanding the underlying principles. The distinction, used here ('written/verbal'), is intended to be 'loose'. This is designed to ensure that the reader will not be forced into exaggerating minor conceptual differences, when it comes to organising this information in their own mind.

Non-Participant Observations

As the name suggests, non-participant observation involves observing others, but not interacting with them. Non-participant observation is another way in which the principles of experimental research can be adapted to a different type of research question. The type of research question to which this method is suited, is one where it is hypothesised that two (or more) groups of people will differ with respect to certain behaviour. For example: boys will play with gender 'appropriate' toys more than girls; men will interrupt more than women, in mixed group conversations; more older than younger people will object to having to wait in a queue. One is still hypothesising that a change in one variable (group membership) will lead to a change in another variable (the behaviour). The only difference is that the (observed) behaviour happens naturally, rather than in response to a situation/stimulus manufactured by the researcher. This does not, however, preclude the possibility that the research may take place in a 'laboratory'. The nature of the 'target' behaviour, itself, gives a good indication of whether a 'lab' based design, or a 'field' based design, is more appropriate/practical. While the behaviour, elicited in observational research, is more 'naturalistic', recording it is more problematical (relative to an experiment). This is true, irrespective of whether the research is being conducted in a laboratory or in a more natural environment. Given that these (practical) problems have to be addressed (in any observational research), it seems only sensible to look at what they are.

The first of these is identifying the 'target' behaviour. This is not as straight forward as it first appears. The following table (overleaf) contains a typology of overt human behaviour. (This typology is taken from the contents of Desmond Morris' book *Manwatching*.)

A Typology of Human Behaviour	
Behaviour	**Explanation**
Inborn Actions	Actions we do not have to learn
Discovered Actions	Actions we discover for ourselves
Absorbed Actions	Actions we acquire from our companions
Trained Actions	Actions we have to be taught
Mixed Actions	Actions acquired in several ways
Incidental Gestures	Mechanical actions with secondary messages
Expressive Gestures	Biological gestures of the kind we share with other animals
Mimic Gestures	Gestures which transmit signals by imitation
Schematic Gestures	Imitations that become abbreviated or abridged
Multimessage Gestures	Gestures that have many meanings
Gestural Alternatives	Different gestures that transmit the same signal
Symbolic Gestures	Gestures which represent moods and ideas
Aesthetic Behaviour	Our reaction to the beautiful in art and nature
Shortfall Signals	When we under-react despite ourselves
Overkill Signals	When we over-react
Contradictory Signals	Giving two conflicting signals at the same time
Intention Movements	Get-ready actions that signal future intentions
Non-Verbal Leakage	Clues that give us away without us knowing
Pupil Signals	Pupil dilation and constriction indicating mood
Autonomic Signals	Actions and other changes resulting from physical stress
Cut-Off	Actions that block in-coming visual signals when we are under stress
Displacement Activities	Agitated fill-in actions performed during periods of tension
Re-Motivating Gestures	Actions which stimulate a new mood as a way of eliminating an old one
Yes/No Signals	Indicators of agreement & acceptance, denial & refusal
Territorial Behaviour	The defence of an area

Barrier Signals	Body-defence actions in social situations
Threat Signals	Attempts to intimidate without coming to blows
Insult Signals	Sneers and snubs – the way we show disrespect
Obscene Signals	The symbolism of sexual insults
Protective Behaviour	Reactions to danger – both real and imaginary
Submissive Behaviour	How we appease our critics or attackers
Religious Displays	Acts and displays of religious deference
Fighting Behaviour	Human combat behaviour
Sporting Behaviour	A modern hunting ritual
Triumph Displays	How winners celebrate and losers react
Status Displays	Ways in which we signal our position in society
Altruistic Behaviour	How we help others at our own expense
Play Patterns	Play signals, play rules and playfulness
Gender Signals	Signals that help to label or emphasise the sex of the signaller
Auto-Contact	Self-intimacies – why and how we touch ourselves
Sexual Signals	The courtship and pre-copulatory behaviour of human beings
Gaze Behaviour	Staring and glancing – the way we look at one another
Tie-Signs	Signals that display personal bonds to others
Body-Contact Tie-Signs	The way companions touch each other in public
Postural Echo	The way people unconsciously act in unison
Taboo Zones	Regions of the human body that are out of bounds
Overexposed Signals	Going too far – breaking through the etiquette barrier
Clothing Signals	Clothing as displays, comfort and modesty
Body Adornment	Social mutilations and cosmetic decorations
Body Self-Mimicry	Ways in which we imitate ourselves anatomically

Parental Signals	Maternal and paternal messages of loving care and safety
Infantile Signals	Baby signals –crying, smiling and laughing
Animal Contacts	Human involvement with other species
Guide Signs	Pointing and beckoning – how we show the way
Redirected Activities	Actions diverted on to a bystander
Baton Signals	Actions that emphasise the rhythm of words
Laterality	Left-handed versus right-handed
Locomotion	The twenty basic ways of moving from place to place
Aquatic Behaviour	How humans behave in water
Resting Behaviour	The postures of relaxation
Feeding Behaviour	How, where and what we eat and drink
Gesture Variants	Personal or local variants based on gestural themes
Regional Signals	Signals that differ by country or region
Hybrid Gestures	Signals made up of two original gestures
Compound Gestures	Actions made up of a number of distinct elements
Technical Gestures	Gestures used by specialist minorities
Coded Gestures	Sign-language based on a formal system
Relic Gestures	Gestures that survive beyond their primary context
Metasignals	What a signal says about other signals
Supernormal Stimuli	The creation of stimuli stronger than their natural equivalents

After Morris, 1980:5-7

While Morris' book provides clear illustrations of each type of behaviour, one should bear in mind that these illustrations have probably been selected, precisely *because* they are clear illustrations of the behaviour. Even so, not all of these behaviours lend themselves to being (dependent) 'variables'. Some are simply too uncommon, or too elusive, for the research to be feasible. Others may be difficult to operationally define, or to define with the degree of precision necessary for an empirical investigation. Alternatively, the degree of precision required may be possible, but the definition may be too complex for use in the field. As mentioned, operational definitions are important because they help to ensure that the two (or more) researchers are looking for the same thing. It has long been acknowledged that what a researcher *looks for* determines what they see. This brings us on to yet another (practical?) drawback of observational research – the problems associated with looking for a certain type of behaviour.

The research question, together with the influences that have shaped that question (see Taylor, 2007), will determine what a researcher is looking for. *'Looking for'* implies that the researcher has certain expectations about what they will see. By implication, this means there are certain behaviours that the researcher is not expecting to see, or even looking for. Let us ignore, for a moment, the more obvious (methodological) influences discussed in my other text, and look at a more fundamental influence – one's culture. What might seem like 'natural' categories, may, in fact, be the product of how one's culture has taught one to 'see', and 'divide up', the world. While this is clearly the case with human behaviour, concepts as basic as colour can be seen to be influenced by one's culture. Anthropologists have found that not all cultures identify the same number of colours, but they do seem to follow a pattern:

> If the language identifies only two colours then these will be 'black' and 'white'.
> If it identifies three colours then the third will be 'red'.
> If it identifies four colours then the next will be 'yellow' or 'green'.
> If it identifies five colours then it will have terms for both 'yellow' and 'green'.
> If it identifies six colours then the sixth term will be 'blue'.
> If it identifies seven colours then the seventh will be 'brown'.
> If it identifies eight or more then the additional four will all, or a combination of, 'purple', 'pink', 'grey' and 'orange'
>
> ***After Berlin and Kay, 1969; 1991 edition:2-3***

This does not mean that people who come from a culture which identifies only two colours cannot perceive a greater number of distinctions; it simply means that their culture places limitations on the terms available to them, when it comes to describing what they can see. Similarly, there are distinctions made by other cultures which are not recognised by our own. For example, *Tagalog*, spoken by people in the Philippines, has no single expression for 'brown'. The word, which is used in its place, depends on the context. To describe food, which has been cooked until 'golden brown', the *Tagalog* speakers use the word *pula*, meaning *'red'*. When it comes to describing the colour of their skin, they use the word *kayumanggi*. For everything else, they use the words *kulay kape*, meaning coffee-coloured, or *kulay-tsokolate*, meaning chocolate-coloured (reported in Berlin and Kay, 1991:100). A more profound example can be seen in the way in which animals were classified (in England) during medieval times. At the time, they had a system of classification known as 'The Great Chain of Being', which classified animals according to their 'spiritual perfection' (see Leahey, 1987:148). To us, this is a cultural, rather than a biological distinction; but to someone from different culture, our 'biological' distinctions might seem 'artificial', even 'unnatural'. Researchers should not, therefore, deceive themselves into thinking that their system of classification is 'natural', let alone universal.

The problem is not just with 'seeing' categories of behaviour, but with 'seeing' the associations between them. In other words, between certain types of behaviours and certain groups of individuals. We can see, in the work of William Sheldon, how a researcher's expectation can lead them to see associations which may not be there. (Or, at least, may not as strong as they first appear.) Sheldon was interested in finding out whether a person's temperament could be predicted from their body type. He found very high correlations between the following body types and temperaments: *Endomorphic* and *Viscerotonic, Mesomorphic* and *Somatotonic, Ectomorphic* and *Cerebrotonic* (Pitts, 1987). Researchers who have attempted to replicate this work, have not been able to achieve such high correlations (between these pairs of descriptors). The explanation, most often put forward, is that (unlike subsequent researchers) Sheldon did both sets of scoring (body type and temperament) himself. Consequently, he may have been seeing what he wanted to see (Pitts, 1987). Again, this is a lesson on how one's expectations can influence what one 'sees'. This is why it is important to have some measure of reliability (usually another observer who can agree the findings).

Even if the behaviour can be operationally defined, and the observers can agree on what they have seen, there are still practical problems associated with observing people. The most obvious question is whether or not one can actually observe the behaviour. If the behaviour is verbal and it occurs in a noisy environment, then observing it may be out of the question. Similarly, if the environment does not lend itself to (the observers) being able to observe the behaviour (without making it obvious), this again may be impractical. If the research involves audio or video taping, then one needs to check that the equipment is working correctly. Less obvious, is that there is limit to the number of behaviours one (an observer) can keep track of. In many respects, the (practical) problems associated with observational research are so obvious that they are often overlooked. There is not much point in operationally defining the variables, if they can not be observed in the field!

To these practical problems, one can add (ethical) issues of (the 'participant's') consent and permission. In this case, the relevant factor is, perhaps, the degree of intrusion. Most (non-participant) observational research is innocuous, because, by definition, its only about overt behaviour. Non-participant observation is not about the 'participants' as people; it is only about their behaviour. As ironic as it may seem, there are probably situations where people would not mind being observed (if it helps an academic understand the situation better), provided that they did not know, or were not told, they were being observed. It is difficult to imagine a researcher going up to a couple in a bar and saying: *"Excuse me, I am conducting research into flirting behaviour. Do you mind me asking how you know each other? I have noticed that...."* Although this is a rather light-hearted illustration, it does bring out one of the main methodological contradictions of this type of design. Aside from the possibility that it (the knowledge that one is being observed) could affect the participant's behaviour, it is not clear what *'the right thing to do'* is, in this situation. What is 'good practice' (i.e. obtaining participant's consent) may actually cause the greatest harm. This is even more of an issue when an observer interacts with the participants – as is the case with *participant* observation.

Participant Observations & Ethnographies

Technically, the difference between a participant observation and an ethnography is one of 'observing' versus 'living as'. Participant observation involves observing people by interacting with them, whereas ethnography involves reporting on what it is like to live as [subject]. While many ethnographers describe what they do as 'participant observation', ethnography should be more than just participant observation. 'To observe' and 'to live as' are qualitatively different activities. In my study of workplace celebrations, I observed, and participated in, a departmental sports' day and barbecue, but I did not experience what it was like to be a member of that department. Some situations can only be observed as a participant, but this is not the same as knowing what it is like to be a member of that social group. In short, participant observations are used to investigate social situations (while participating in them); whereas ethnographies are used to investigate social groups (from an insider's perspective. The difference between a participant observation and a case study (discussed in chapter five) is, essentially, one of 'situation' versus 'context'. In my study into workplace celebrations, the research question required me to observe *an* informal workplace celebration, as part of an investigation into these types of *events*. Given its focus, the work of the department, its role and function within the organisation, its history, and its performance, were not relevant to the investigation. Even the structure of its staff, and their working relationships, were not *that* relevant. All of these would, of course, be relevant if this was a case study (of the department).

The difference between an ethnography and a case study is one of 'reporting as' versus 'reporting on'. The terms 'ethnography' and 'case study' are often used quite loosely: if it is an ethnography, then it should be about the researcher's own personal experience of what it is like to *be* that subject. Otherwise it is a case study (if it is good enough to be called that).

Participant Observations

The difference between participant observation and non-participant observation is more than just a matter of interacting with those being observed. Participating means more than just being *there*; it means taking an active role. The degree to which a person can successfully take on this role (of 'participant'), varies depending on the person and the nature of the situation. In this respect, it is just like any other social role. However, when the role (of participant) is part of a research design, the researcher's ability to adopt this role becomes central to the success of the research. What a researcher will 'see', will depend on how well they can carry-off this role. If they can carry it off successfully, then they will see the participants behaving as they would with another participant; if they can not, then they will see the participants behaving as they would with an 'intruder'. Clearly, some people will be able to adopt some roles better than others; it is not an either/or situation, but a matter of degree. As obvious as this may sound, there are a surprising number of researchers who seem to believe that obvious differences in race, age, and social status will be overlooked by participants, and that they will be treated 'just like everyone else'.

Elliot Liebow was a researcher who thought that he could use participant observation to learn about black, street corner, men. As a white man, this is what he learned:

> My field notes contain a record of what I saw when I looked at Tally, Richard, Sea Cat and the others. I have only a small notion – and one that I myself consider suspect – of what they saw when they looked at me.
>
> Some things, however, are very clear. They saw, first of all, a white man. In my opinion, this brute fact of color, as they understood it in their experience and as I understood it in mine irrevocably; and absolutely relegated me to the status of outsider. I am not certain, but I have a hunch that they were more continuously aware of the color difference than I was. When four of us sat around a kitchen table, for example, I saw three Negroes; each of them saw two Negroes and a white man....
>
> I used to play with the idea that maybe I wasn't as much of an outsider as I thought. Other events, and later readings of the field materials, have disabused me of this particular touch of vanity.
>
> Whenever the fact of my being white was openly introduced, it pointed up the distance between me and the other person, even when the intent of introducing it was, I believe, to narrow that distance.
>
> ***Liebow, 1967:248-9***

Liebow was willing to accept that he could not overcome the racial difference, but other researchers claim to have overcome far more significant differences (between themselves and the other participants). Here are two quotes, from two different researches, who clearly believe that they were able (eventually) to blend in to their setting. In this case, the setting was either an infant school (three to six year old children) or a primary school (seven to ten year old children) classroom, and the role was that of a pupil!

> I rapidly learnt that children in infants' classrooms defined any adult as another teacher or teacher surrogate. To avoid being engaged in conversation, being asked to spell words or admire pictures, I evolved the following technique. To begin with I kept standing so that physical height created social distance. (I noticed how the teacher often got down to their level.) Next, I did not show immediate interest in what the children were doing, or talk to them. When I was talked to I smiled politely and if necessary I referred the child asking a question to the teacher. Most importantly, I avoided eye contact; if you do not look you will not be seen.
>
> ***King, 1978:4***

> Though it may seem incongruous to think of a middle-aged, 200-pound male anthropologist being a participant-observer in third- and fourth-grade classes, this was actually the case. I sat at a desk in the back of the room and did the same things the children did insofar as my ethnographic recording activities permitted. The children accepted me and my role much more quickly than did the teachers, but both seemed to adapt to the incongruities after a period of several weeks. I can think of no other way that I would have come to an understanding of what the third and fourth grades in the [school]...were like.
>
> ***Spindler, 1974:385***

The extent to which either of these researchers *'participated'* in this research setting is perhaps questionable. If we look at the procedures, the first investigation can be more accurately described as a *non*-participant design. The second investigation is described (by the researcher) as "ethnographic", which means that the researcher was not just observing the situation, he was actually experiencing what it was like to be a pupil in that situation. While 'common sense' has long been rejected as means of support for academic arguments, there is something intuitively naïve about this design. Experiencing the situation as a 'participant' depends on more than just being 'accepted' (as a 'participant'); it depends on being able to see the situation as the *other* participants see it. While we can accept that this researcher was able to take on the role of 'a participant', it is difficult to accept that he was able to take on the role of 'a schoolchild' (which is what he would have had to have done, as an ethnographer). Even as a successful 'participant', he was not observing a class of children, but a class of children with another (himself) adult present. Both researchers acknowledge that it was some time before the other participants (the children and the teachers) accepted them. This stage of the process can perhaps, more accurately, be described as a period of adjustment. Researcher can not afford to assume they are not effecting (or have not effected) the behaviour of their participants. In these examples it is obvious, but just because it is not obvious does not mean to say that it is not happening. The best a researcher can expect is to affect the group no more than a 'natural' participant.

One of the reasons why so many researchers have difficulty fitting into their research setting is because they are not usually the kind of situations in which an academic would (normally) find themselves. It, therefore, comes as no surprise that researchers often have difficulty gaining access, and fitting in, to such settings. Gaining access (to the research setting) is, therefore, seen as an important part of the process, not just because it allows the researcher to conduct the research, but because it allows the researcher to communicate what that was like, for them, to enter that kind of situation. It should, of course, be remembered that this is the experience of gaining entry for someone who would not normally go into that situation. Someone, who is better able to relate to the people in that situation, may experience the process differently. This suggests that it may be better for a researcher to investigate a situation with which they are already familiar. However, this would not make much of an 'investigation'. A case in point is Mario Brajuha's study of restaurant work. Most people know what it is like to be employed in this kind of work, so it can hardly be described as a study of 'alien cultures'. Arguably, the point of conducting this type of research is to investigate the lives of people who are normally 'inaccessible'. The researcher could, of course, employ people (as 'observers') who are already participants in the setting (but they would not be trained researchers). While, both of these approaches raise the issue of objectivity (being able to look at the situation without any preconceptions), this is only 'an issue' for those who have certain beliefs about the nature of knowledge. The more the (research) design resembles an experiment; the more likely it is to be influenced by Positivist beliefs (about the nature of knowledge). Similarly, the more the (research) design resembles an investigation into lived experiences; the more likely it is to be influenced by Feminist beliefs about the nature of knowledge. (Participant observation and non-participant observation are among those influenced by both.) As Positivism gives way to Feminism (in a research design), objectivity gives way to reflexivity.

The researcher's ability to adopt a 'participant' role (in that particular research setting), also has implications for their ability to maintain control over the research (data). If a researcher has to have someone to 'show them around', there is a danger that the 'findings' will simply reflect what the researcher has learned from this person, rather than what they have learned from their observations. While there no rules as to how much a researcher *has* to 'participate', loosing control of the research, to this extent, must call into question the validity of the research. The research purports to be the observations of a well trained researcher. While this may be true (as far as it goes), the observations are not under the *direction* of the researcher. Given that the research is conducted under the auspices of the researcher's expertise, one (the reader/academic community/public) might reasonably assume that it is being conducted under the direction of the researcher (and clearly this is not the case). In situations where another person has the researcher at a (permanent) disadvantage, the researcher is likely to end up just being *led* by that person (as Polsky found out in his study of criminal behaviour):

> Studying a criminal in his natural setting means not only studying him outside of any law-enforcement context. It means studying him in his usual environments rather than yours, in his living quarters or streets or taverns or whatever, not in your home or your office or your laboratory. And it means you mustn't "schedule" him, mustn't try to influence his shifting choices among his environments or interfere with his desire for mobility or immobility. If he wants to sit in front of his TV set and drink beer and watch a ballgame for a couple of hours, so do you; if he wants to walk the streets or go bar-hopping, so do you; if he wants to go to the racetrack, so do you; if he indicates (for whatever reason) that it's time for you to get lost, you get lost.
>
> ***Polsky, 1967:135***

If this is the kind of relationship that a researcher has with other participant(s), what has been achieved by using an observational design (and a participant one at that)? One of the reasons for using an observational design is to reduce the possibility of the information being distorted by a third party. (The researcher is able to see the situation for themselves.) If a researcher is going to allow other participants to dominate the research, in this way, then there is little point in using a direct observational design. Moreover, then they can not (personally) guarantee the quality of the evidence. Researchers need to look carefully at the level of access they have (been given) to a particular situation, and whether participant observation will be effective. Peter Donnelly's (1989) study of erotic movie theatres is a good example. An erotic movie theatre is the kind of place where it is possible to use *participant* observation effectively. The nature of situation means that some participants prefer to remain alone and anonymous. It is, therefore, possible for a researcher to *participate*, as one of these 'alones' (ibid:245). In this kind of situation, the researcher does not have to rely on others, in order to play their role. There are, of course, times when this can be a disadvantage – as Peter Donnelly recalls:

> The appearance of acquaintances (five) at Cinema A on separate occasions was managed in different ways. On the first occasion, seeing four students, I felt that any attempt to explain my reasons for being there would have been met with

> knowing looks on [and?] comments, and possibly eventual rumors. I resolved the problem by literally hiding, the first time behind a pillar in the lobby, the second time in the toilet, and on the third and fourth occasions I went to take my seat in the relative darkness of the theatre a little earlier than planned. On the fifth occasion a female colleague appeared in the company of several other people. After a moment of panic and indecision about whether to hide again, I decided to reveal myself and my purpose and conduct an impromptu interview. She claimed never to have seen a pornographic film before, was curious, and concluded by asking me what she should expect. [At this point, the report changes subject.]
>
> ***Donnelly, 1989:247***

The alternative approach is, of course, to observe situations in which one is already a participant. While this may eliminate the problem of not being able to adopt a 'participant' role (successfully), it may mean deceiving one's friends. The issue is not so much: *'who to tell and who not to tell'*, but: *'who to tell, how much, and when'*, and this all depends on the researcher's relationship (and prospective relationship) with other participants. A case in point is that of Mario Brajuha:

The Mario Brajuha Case

Mario was employed as a waiter in a 'fancy' restaurant and decided to make this (the social interactions of the restaurant) the subject of his Ph.D. Eighteen months later, the restaurant burnt down – under suspicious circumstances. In the course of their enquiries, the police learned that Mario had made notes on various conversations which had taken place and they wanted to see his notes. Mario refused:

> Protecting my notes became a major preoccupation. I thought: what if somebody reads this? There's lots of stupid stuff, conversations with my wife, personal jokes, mistakes. I was embarrassed at the idea and was afraid it might lack profound sociological content and that it would be showing my whole private life.
>
> ***Brajuha and Hallowell, 1986:457***

Unfortunately, the grand jury did not see it that way, and Mario was subpoenaed to produce his notes in court. Mario resisted, against the advice of the university attorney, seeking legal representation elsewhere. At first, no one would defend him. Mario was, therefore, faced with the choice of "jail for contempt or surrender the notes" (ibid:458). A stay of execution, came from a willing attorney and a quirk of New York legislation, which stated that the law offered protection to "privileged communications by those engaged in disseminating knowledge to the public" (ibid.). The case was adjourned, while the judge considered whether it applied to Mario's notes. At this point, the *Assistant District Attorney* was willing to make a deal with Mario – the withdrawal of the subpoena, in exchange for a little investigating on his behalf. Mario turned it down. It was eventually decided that the notes were *not* 'protected' by the New York legislation.

According to Mario, the owners used the delay as an excuse to not re-open the restaurant. Consequently, some of the staff had to find (less well paid) employment elsewhere. Those who were left, seemed to dislike Mario for some reason:

> I was in the kitchen and [the owner] came from behind me and hit me on the head with the serving fork…Many times before he hit, kicked, spilled hot water, or kicked me in the ass. Sometimes he would come behind me and lock the door while I was in the refrigerator.
>
> ***Field notes, cited in Brajuha and Hallowell, 1986:457***

Mario's problems were not only confined to the kitchen. He soon learned that "other parties" (ibid: 460) were interested in his notes. This prompted him to hide the notes in a colleague's study. A few days later, a colleague discovered a "burly, unfamiliar young man rummaging around his office" (ibid). At this point, the case began to take its toll on Mario's personal relationships. However, the case ended happily, with a number of academics rallying to his support.

Mario Brajuha had more problems than most, when it came to deciding who to tell (who not to tell), how much, and when. His decisions ended up putting quite a strain on his relationships with: the other restaurant workers, the former restaurant workers (who lost their jobs because of him), the restaurant owner, the police, the law courts, his (academic) colleagues, his family, and his friends. This raises a very important question (from a researcher's point of view): *'how many people's trust would have to be betrayed, in order to get the information for the research, and would it be worth it'?* While this may not look like much of a problem, many studies betray the researcher's feelings of guilt, at having used such a design. Ronai and Ellis's (1989) study of table dancing is a good example. Ronai deceived her former colleagues, but she justified it by saying that deception is endemic to this kind of environment (a table dancing club). Likewise, she justified using people (observing them for her research) by saying that the girls were being used by the men (for turn-ons). Therefore, it was all right for them (her and her co-researcher) to use the men (as the subject for her research). Are the researchers trying to convince us, or are they trying to convince themselves? The research design is entirely appropriate, they do not need to use these kinds of justifications.

Many of the problems, associated with participant observation, stem from the need to perform two different roles simultaneously (participant and researcher/observer). What makes this type of design taxing (on the researcher), is not just the need to perform two different roles, but the *demands* of those roles. It is difficult for a researcher to perform both roles effectively: either the quality of the data suffers, or the researcher's 'performance' suffers. The method also presents a researcher with more personal challenges: they either have to deal with an alien 'environment', or the consequences of alienating others. Although these issues have been raised in the context of an observational design, many of them are relevant to ethnographic research. The major difference, between an ethnography and a participant observation (in terms of design decisions), is that an ethnography is rarely used for the study of a group to which the researcher already belongs. An ethnographic design requires the researcher to be able to reflect on their experiences, as a 'native'/ participant, from the perspective of a researcher. If a researcher can not stand back from their 'native' role, then the investigation will simply not succeed.

Ethnographies

Ethnography takes the idea of investigating through participating one step further. In this case, the researcher is not simply observing while participating, but reflecting on their experiences of *living as* their subject. On the face of it, they may not appear to be all that different. In this case, the difference is conceptual rather than procedural (living as a participant, rather than observing as a participant). It does not help when researchers use the terms ('participant observation' and 'ethnography') interchangeably. (As some do.) Similarly, there can often be little difference between a *poor* ethnography and a *good* participant observation. An ethnography *should* communicate something of what it feels like to be a member of that (social) group, and how a member of that (social) group sees things. Given the complexities of the design/idea, it is difficult to advise someone how to go about conducting an ethnography. There are, however, a number of ('how to') texts, which detail the practicalities, these include: *Ethnography Principles in Practice* by Martyn Hammersley and Paul Atkinson (1992), *The Ethnographic Interview* by James Spradley (1979) and *The Professional Stranger* by Michael Agar (1980). (A much wider review of the same can be found in Hammersley and Atkinson, on pages 238 to 245.) Given that this book is intended to be a discussion of the rationale, rather than the mechanics, of social research, the following section will aim to guide the reader through the various texts on the subject of ethnographic research.

The first thing to say about these texts is how tightly structured they are. In this case, the structure is not simply that of the book, but that of an ethnographic investigation. The way in which the stages of the investigation are delineated, in these texts, is at odds with the (inter)personal nature of the method. Arguably, the method should be described in a way, which reflects the impromptu nature of the design. This apparent inconsistency begs the question: *'why should such a spontaneous method be described in such a prescribed way?'* We can take this one stage further and ask: *'what does this say about the author's view of knowledge, and the knowledge creation process?'* A tightly prescribed ('checklist') approach (to research) suggests that the author is leaning towards the idea of a standardised procedure, and this desire for standardisation is an attribute of Positivism. This can clearly be seen by looking at the beliefs, about knowledge, which underlie a standardised procedure. A standardised procedure is the first step towards making the knowledge gathering process independent of the researcher. (Turning the researcher into a tool of the research process/science.) This would make the knowledge independent of any human 'knower', and this belief, that knowledge is independent of any human knower, is Positivistic (see Taylor 2007 for a discussion of methodologies). Contrary to how it may appear, these authors are *not* Positivists. They are not arguing *for* Positivism; they are arguing *with* Positivism (and the Positivists). Many of them seem to be trying to convince an imaginary Positivist that ethnography can be 'scientific. It is important to be able to separate out these methodological arguments, because this allows one to see the features of the method more clearly. The approach taken, in this book, is that ethnography can be used with a number of methodologies. Therefore, allowing oneself to be drawn into these (methodological) debates is tantamount to accepting the tenets of Positivism.

Many of the texts (on ethnography), written by Martyn Hammersley, show a preoccupation with Positivist methodology. Although Hammersley's books are a defence of ethnography (as a 'scientific' method), they are a defence against an imaginary Positivist criticism. In defending ethnography against Positivism, Hammersley is reinforcing the idea that Positivism *is* the methodology which defines 'what is scientific' in academic research. By accepting that he needs to argue his case in Positivist terms, Hammersley is, effectively, accepting that Positivism is dominant, if not the defining, methodology. (Or, as Foucault would say, the discourse is still being dominated/controlled by Positivism/Positivists, therefore they have the power.) By casting Positivism in the role of 'the one to defeat'/'the one to placate', Hammersley is effectively strengthening its status as a methodology (which is the very opposite of what he sets out to achieve). In his book *What's Wrong With Ethnography? Methodological Explorations* (1992) there is a chapter (number four) entitled *'By What Criteria Should Ethnographic Research Be Judged?'*; in that chapter he says: "(w)hile the concepts of validity and reliability, suitably defined, are appropriate criteria for judging the methods and products of research, they are not sufficient" (1992:67), he then goes on to supply his own supplementary criteria. The problem with using a standardised criteria is that it is methodologically inconsistent with the method itself. The ethnographic method is, by its very nature, idiographic (individualistic), whereas the idea of a standardised criteria is, by definition, nomothetic (and this seems methodologically inconsistent). Once again, the idea of a standardisation (a standardised criteria) is a Positivistic belief. What Hammersley should have done (if he had wanted to win the philosophical argument) was to argue that it is inappropriate to assess ethnography in this way.

Positivism has become so much a part of (the discourse of) academic research that many academics think they are being 'scientific' when they are actually being 'Positivist'. As far as most academics are concerned, this is what it means to be 'scientific'. Hammersley obviously believes that his work undermines Positivism, and Positivist beliefs about knowledge; what he does not realise is that he is using Positivism to do it. Take the following quote (from Hammersley and Atkinson):

> Once one begins to recognize the complexity of the scientific enterprise, the different functions that research can serve, and the failings of the single study model, one is in a better position to appreciate the contribution that ethnography can make to social science. It should be clear that we do not regard ethnography as an 'alternative paradigm' to experimental, survey, or documentary research. Rather it is simply one method with characteristic advantages and disadvantages, albeit one whose virtues have been seriously underestimated by many social researchers owing to the influence of Positivism.
>
> ***Hammersley and Atkinson, 1992:23***

The word 'paradigm' is usually used to mean a school of thought or a theory. To describe ethnography as an 'alternative paradigm' (even if it is to negate the idea) implies that it is a different way of thinking about academic research. Perhaps what they mean to say is that its underlying methodology is believed (by other academics) to be different to that underlying the other methods. They then go on to criticise the Positivist nature of social research, suggesting that this is why ethnography is not seen as 'good' academic research. (Presumably, because it does not conform to the ideals of the Positivist approach.) Again, Hammersley's position is not very clear. On the one hand, he (they) seem/s to be arguing that all methods should be seen as equal, but on the other, he (they) is/are opposed to a single dominant methodology/philosophy. The phrase "we do not regard ethnography as an 'alternative paradigm' to experimental, survey, or documentary research" implies that they see these methods as part of the same 'paradigm'. If all of these methods are to be thought of as the 'same', then what school of thought (paradigm) sees them as such? In this case, the word 'paradigm' seems to mean just 'social research'. If this is the case, then why not say so? An ethnography is *not* the same as an experiment, a survey, or a piece of documentary research; they are all different methods, suited to different research questions, and different methodologies. Given the notoriety of the Chicago school, it is questionable whether ethnography has ever been regarded as inferior to these other methods anyway.

Ethnographic research is considerably more difficult to manage than any of the methods described above. As mentioned, it requires the researcher to adopt the role of a 'native' (a member of the social group they are investigating). There are a number of issues and problems specific to this type of research activity, not least of which is the need to gain **access** to the 'field'. In order for this type of research to be effective, the researcher has to gain access on at least two levels: *physical* (access to the setting itself) and social (their accepted by the group). Physical access is often controlled by individuals, who are in a position to give the researcher entry to the field. They are often referred to as a **'gatekeepers'**. The following quotes illustrate not only how gatekeepers can play a pivotal role in ethnographic research, but that the role of gatekeeper itself does not require any *formal* authority. The first (from Whitten) is an account of what happened when he tried to enter a small rural community in Nova Scotia:

> My phone conversation with ministers went well, and I was encouraged to call the councilor…I introduced myself as an anthropologist from the United States, interested in problems encountered by people in rural communities in different parts of the Americas. Following procedures common in the United States and supported by educated Nova Scotians, I said that I was particularly interested in Negro communities kept somewhat outside of the larger social and economic system. I was told, politely, but firmly, that the people of the rural Dartmouth region had had enough of outsiders who insulted and hurt them under the guise of research, that the people of the region were as human as I, and that I might turn my attention to other communities in the province. I was asked why I chose "Negroes" and when I explained that Negroes, more than others, had been excluded from full participation, I was again told that the people of rural Nova Scotia were all alike, and that the colored people were tired of being regarded as somehow different, because there was no difference.
>
> ***Whitten, 1970:371***

Whitten's mistake was to not investigate the local protocols thoroughly enough. His first mistake was to approach the official directly. The norm, in this situation, is to have a native (who can vouch for your character) introduce you to the official. His second mistake was to demonstrate an awareness of certain ethnic distinctions. Nova Scotians expect to tell outsiders what (ethnic) distinctions exist in their community, not the other way around (ibid.). William Foote Whyte was considerably more fortunate when it came to finding someone who could give him access to his 'field' *(Street Corner Society).* The person who gave him access was a gang leader, by the name of 'Doc'. As this quote (from 'Doc') shows, he was a very useful person to know:

> "Well, any nights you want to see anything, I'll take you around. I can take you to the joints – gambling joints – I can take you around to the street corners. Just remember that you're my friend. That's all they need to know. I know these places, and if I tell them that you're my friend, nobody will bother you. You just tell me what you want to see, and we'll arrange it."
>
> ***'Doc' in Whyte, 1955:291, from the 1993 edition***

The next step, for the researcher, is to find a role that will allow them to live in the field (but not necessarily as a member of that social group). This is more likely to be one which is acceptable to (and chosen by) the group, rather than one which the researcher chooses for themselves, a point noted by Agar:

> When you begin doing ethnography, group members are going to wonder who you are. They will listen to you and watch your behavior, and they will draw on their own repertoire of social categories to find one that fits you.
>
> ***Agar, 1980:54-5***

An good example, of this, can be seen in Hansen's study of Catalonian village life:

> Initially the interviewing process went very slowly because I was overly polite and solicitous about seeking interviews with people I hardly knew. I made the error of being too formal, which made these people suspicious of me. My mistake was brought home to me forcefully by one of the few nobles remaining in the Alto Panades, whom I had interviewed by chance. He explained in no uncertain terms that I was behaving like a servant or client to these individuals when my own wealth, looks, and education meant that I was superior to them. Therefore, he said, I should command them to give me interviews. He proceeded to accompany me to more than twenty bourgeois landholders, and ordered them to give me what I wanted, on the spot, including details of business scandals, etc. All complied, some with obeisance toward the count, and all with both deference and expansiveness toward me. The count checked all their answers to see if they were concealing vital information. Astonished and embarrassed as I was, the count had a point: After these twenty interviews, I was swamped by volunteers. It had suddenly become fashionable to be interviewed by *el distinguido antropologo norteamericano.*
>
> ***Hansen, 1977:163-4***

Once a researcher has been assigned a (native) role, they then have to maintain that role. This is essential for the success of the research. If a researcher can not maintain their (native) role, then they will not be able to remain in the 'field', and they will not be able to continue with the research. It is likely to be the case that there will be certain individuals who have helped the researcher to gain access to the field, and/or are aware of the researcher's dual role. A researcher, therefore, has to keep on good terms with these people, in order to remain in the field. The following (quote) shows the lengths that Patricia Adler, and her future husband, had to go to in order to maintain their roles (which were not dealers) in a small-time drug dealing community.

> In our case, we offered everything that friendship could entail. We did routine favors for them in the course of our everyday lives, offered them insights and advice about their lives from the perspective of a more respectable position, wrote letters on their behalf to the authorities when they got in trouble, testified as character witnesses at their non-drug-related trials, and loaned them money when they were down and out. When Dave (one of their informants) was arrested and brought to trial for check-kiting, we helped Jean (his wife) organize his defense and raise the money to pay his fines. We spelled her in taking care of the children so that she could work on his behalf. When he was eventually sent to state prison we maintained close ties with her and discussed our mutual efforts to buoy Dave up and secure his release. We also visited him in jail.
>
> ***Patricia Adler, 1985:16***

While (technically) this is only a participant observation, at appears that at least one of their colleagues took the research a stage further. The individual is described as having 'gone native'. ('Gone native' is a phrase used, in ethnography, to describe a researcher who can no longer look at their 'native' role objectively.) As the following quote shows, this was part of a more serious problem:

> (W)e turned down an offer by a friend to help us gain research entrée with agents of the Drug Enforcement Administration (DEA), data that would have given our work a multiperspectival vantage on dealing but that might have brought our research and our subjects to the attention of the authorities. We also feared being subpoenaed to testify in the trial of a former graduate student, whom we had heard through our informants had gone native after being arrested for dealing and he was trying to claim research as his defense. This would also potentially disclose our research and endanger our subjects, we feared.
>
> ***Adler and Adler 1993:251***

These extracts show the lengths to which some researchers have to go, in order to remain in the field. However, ethnographic research is not simply a matter of maintaining one's native role (for as long as necessary), it is a matter of maintaining one's native role, while acting as a researcher. Clearly, this is a strain for the researcher. This next extract shows what can happen when a researcher spends most of their time 'being a native'. It comes from a study of teaching, by Everhart:

> (S)aturation, fieldwork fatigue, and just plain fitting in too well culminated…in a diminishing of my critical perspective. I began to notice that events were escaping me, the significance of which I did not realize until later. For example, previously I had recorded in minute detail the discussions teachers had on categorizing students and those conversations students had on labeling other students. While these discussions continued and were especially rich because of the factors that caused these perspectives to shift, I found myself, toward the end of the study, tuning out such discussions because I felt I had heard them all before when, actually, many dealt with dimensions I had never considered. On the one hand I was angry at myself for not recording and analyzing the category systems, on the other hand I was tired and found it more natural to sit with teachers and engage in small talk. The inquisitiveness had been drained from me, and I could do little to restore it.
>
> ***Everhart, 1977:13***

There comes a point when a researcher has to decide whether to give priority to their native role or their role as a researcher. Devoting more time to the assigned role makes the lived experience more valid, but this means maintaining a deception. Maintaining a deception not only diverts the researcher's attention away from the task of being a researcher (i.e. recording one's experiences), but it can also create ethical problems. Alternatively, a researcher may choose to 'come clean', and relieve themselves of (some of) the burden of maintaining this deception. This allows the researcher to concentrate on the task of gathering data. The researcher is, of course, the person best placed to make such a decision. As well as its timing of any 'disclosures'. The following extract (from Chambliss, 1975) shows how the decision, to 'disclose', can often pay off:

> I first began my research by going to the slums and ghettoes in Seattle. I chose these areas to work in because the official (government) statistics showed them to be the areas with the highest crime rates. In the beginning I had no intention of studying either organized crime or professional theft. Rather, I wanted to investigate the way the police decided to arrest some people and not others...
>
> Within three days I discovered that widespread gambling, prostitution, drug trafficking, and pornography distribution were being completely ignored by the police, while drunks on skid row and Black teenagers in the ghetto were being regularly busted…
>
> Within a week I was convinced that the rackets were highly organized. The problem became one of discovering how, and by whom. I was sitting talking to (the waitress)…when a man I recognized as a policeman came through the door and went into the manager's office. I asked (the waitress)what he was doing:
>
> Waitress:"He's the bag man."
> Chambliss:"The what?"
> Waitress:"The bag man. He collects the payoff for the people downtown."
> Chambliss:"Oh."
>
> I spent the next two months talking informally to people I met at different games, in pornography shops, or on the streets. I soon began to feel that I was at a dead end. By the time I had discovered the broad outlines of organized crime in

> Seattle, but how it worked at a higher level was still a mystery. I decided it was time to "blow my cover."
>
> I asked the manager of the cardroom I played in most to go to lunch with me. I took him to the faculty club at the University of Washington. This time when he saw me I was shaven and wore a shirt and tie. I told him of my "purely scientific" interests and experience and, as best I could, why I had deceived him earlier. He agreed to help.
>
> Soon I began receiving phone calls: "I understand you are interested in Seattle. Did you ever think to check Charles Caroll's brother in law?" And there was one honest-to-God clandestine meeting in a deserted warehouse down at the wharf....
>
> Over the next ten years I pursued this inquiry, widening my contacts and participating in an ever larger variety of rackets. As my interests in these subjects and my reliability as someone who could be trusted spread, I received more offers to "talk" than I had time to pursue.
>
> ***Chambliss, 1975:36-8***

At some point, the fieldwork will 'finish' and the researcher will have to leave the situation. In some cases, there may be nothing more to be gained from remaining in the 'field' (Hammersley and Atkinson, 1992:103-4), but, in others, it may simply be that time and/or money has run out. Few texts offer any advice on *leaving* the field. It may be that they do not see it as *that* important, or *that* difficult to manage. This may be the case, but there again, it may just be the opinion of academics. If everyone (including key informants and other participants) is happy with the way in which the researcher has left the 'field', then there is no problem. If, however, the researcher leaves the field it such a way as to make it look as though they have finished using everyone, and it's time to move on, then the researcher needs to spend time thinking about how they will manage this situation. One of the few researchers to consider this issue (although not in the context of an ethnography) is Sue Cannon in her study of cancer patients. The following (quote) shows just how complicated this can be:

> Long before the fieldwork period ended it became clear to me that I would not be able to completely 'leave the field': that is, I knew that I could not (nor did I want to) say "thank you and goodbye" and tell the women I would not be seeing them again....Liz said "but you won't be finished with me then will you?" (that is after the fieldwork was completed). It became clear that the constraints upon my leaving bore greater significance than those which exist in the case of breaking off a personal friendship. For some women the prospect and fear of abandonment, which was linked in their minds to the point at which medical staff could no longer actively help them and treatment ceased, also applied to the serving of their personal involvement with me and mine with them, and with the study. Some form of compromise needed to be devised. [Sue Cannon devotes the next three pages to saying how she did that.]
>
> ***Cannon, 1989:74***

While ethnography is based on the idea of the researcher living as a 'native', this native existence is seen in the context of an academic investigation. In other words, the ethnography is seen as just a method serving the investigation. The problem with this approach is that the researcher is not actually adhering to the norms and values of the group, they are 'adhering' to the norms and values of the group, because this is what the method requires them to do. In other words, they are not actually adhering to the norms and values of the group, they are adhering to the norms and values of the academic community, which say that they should adhere to the norms and values of the group, when using this method. This is indicative of the attitude that a researcher can simply enter the field, live as a native for a while ('adhering' to the group's norms and values, because this is what their norms and values tell them to do), and then leave. Many researchers seem to (genuinely) believe that the 'natives' will understand their position, and allow them to follow their own norms and values (in other words those of the academic community), rather than those of the native culture (when it serves the interests of the investigation). Why should the 'natives' allow the researcher to follow a different set of norms and values (of the academic community) to everyone else (when it suits them)? It is tantamount to asking the natives to accept that the researcher's norms and values take priority over theirs. This begs the question: *'why should a researcher think this?'* The answer lies in value placed on science and the pursuit of knowledge. In our (and the researcher's) sub-/culture, their value is without question, but this may simply reflect what is valued in *our* sub-/culture. Expecting people from an other sub-/culture to make concessions, on the basis that the researcher's culture values the cause for which they are being made, assumes that the natives accept the norms and values of the researcher's culture, which it necessary for them to accept the norms and values of the researcher's culture!! The illustrations used here (in this section) do not show researchers imposing their norms and values on the natives; they show researchers having to yield to the norms and values of the natives (for the sake of their research). The fact that the researcher has to follow the norms and values of the academic community, as well as the norms and values of the native community, is *their* (the researcher's) problem. Patricia Adler felt it was most unfair that she had to adhere to the norms and values of the academic community, *and* the norms and values of the criminal fraternity, when her criminal informants only needed to adhere to the latter:

> They [the criminals] were allowed to lie to us, borrow money and not repay it, and take advantage of us, but we were at all times expected to behave honorably…It made us feel like we were selling ourselves, our sincerity, and not getting treated right in return.
>
> ***Patricia Adler, 1985:26-7***

Broadly speaking, the ethnographies which seemed to work best (out of the above examples), were those where the researcher was willing to adapt the research (and themselves) to the norms and values of the native culture. (Eventhough it may have meant violating some of their own norms and values.)

Many texts on ethnographic research read as though the researcher has total control over the situation: people are systematically interviewed, documents are systematically analysed, and typologies are systematically constructed (Agar, 1980; Hammersley and Atkinson, 1992; Spradley, 1979). Ethnographies are *not* like experiments or surveys:

they can not be controlled (by the researcher) to the same degree. While the basic design (of an ethnography) may be supplemented with interviews, documents (Agar, 1980; Hammersley and Atkinson, 1992; Spradley, 1979) and even photographs (Prosser, 1992), an ethnography is, in essence, an account of what it is like to be a member of a particular social group. So much of what happens 'in the field' depends on what kind of relationship the researcher has with their *informants*. The researcher's entry into the setting, their role, their level of participation and the data they are witness to/given, are (to an extent) determined by those already in the 'field'. Unlike most of the methods discussed in this chapter, *ethnographic* research requires the researcher to be reactive rather than proactive. Surely, the reactive, interpersonal nature of the fieldwork, would make any attempt to standardise, or control, the design (Positivist principles) futile and inappropriate. So why do academics still try to convince the Positivists that ethnography can accommodate Positivism, when the nature of the method is clearly incompatible with Positivist principles? Again, it comes back to the idea of what is (believed to be) 'scientific', and the unwarranted defensiveness of ethnographers. Social research is made up of a plurality of methods and methodologies, each suited to a (at least one) particular type of research question. Positivism owes its status to history, nothing else.

Conclusion

In this chapter, we have seen why it is important to choose a methodology before choosing a method, and why certain methods may not fit well with certain methodologies (see Taylor 2007 for a discussion). We have also seen how the researcher's role may change depending on the nature of the (research) method employed (and this is a theme, which will continue). The methods, discussed in this chapter, are characterised by the degree to which the researcher can control the data gathering (stage of the research) process. They are the methods which give the researcher the most control over the data; they do not give the researcher *absolute* control over the data. It is also important to realise also that they are the methods, which, *technically*, give the researcher control over the data. As we have seen, collecting the data 'first hand' does not guarantee the researcher the most control over this process. The next three chapters are organised in terms of the degree of control to which the researcher can exercise over the data. Chapter three will look at personal information elicited by the researcher, chapter four will look at personal accounts, and chapter five will look at public records and the media. It may not be apparent from these descriptions, but they represent a progressive decline in the researcher's control over the data. Consequently, they present the researcher with a different set of challenges. Not least is the need to determine the validity (accuracy) of the information.

3: Research Methods and Sources – Researcher Elicited Personal Information

Introduction

The methods, discussed in the previous chapter, allow the researcher to have direct access to the data. All the researcher has to do is create the right experimental conditions, or find the right situation (or people) to observe, and the data is there! However, when the data is personal information about members of the public (preferences, attitudes, opinions, purchasing habits), the researcher does not have direct access to it. The researcher must, therefore, elicit the information from the people who do have direct access to it. In other words, the people who hold these opinions, attitudes, beliefs, etc. Although most of the methods, for doing this, reduce down to simply asking the respondent to answer the relevant questions, the wording of the questions, the mode of the reply, and the context of the inquiry, can all make a difference to the participant's responses. Not only this, but the nature of the research question and the information required, means that some methods will be more appropriate than others.

At this point, it is perhaps worth mentioning that, in this context, 'personal' simply means 'pertaining to the person', it does not necessarily mean that the information is personal (i.e. intimate). In most cases, the information, which the researcher is requesting, is quite innocuous. The reason for eliciting it is to obtain an overall impression of *'society'/'public opinion'* (at whatever level). Although the information comes from individuals, those *individuals* make up 'society'. The methods discussed in this chapter are, therefore, used to investigate things such as (research questions pertaining to): lifestyles of certain groups, responses to advertising campaigns, attitudes towards political decisions, opinions of public services, and so on. In each case, the target population may be as general or as focused as necessary.

Researcher Elicited Personal Written	**Surveys, Researcher Elicited Correspondence & Diaries, Diary Interviews**
Researcher Elicited Personal Verbal	**Interviews and Focus Groups**

Researcher Elicited Personal - Written

In this chapter, the distinction between 'written' and 'verbal' is a reflection of the precision with which the instrument questions can be articulated. If the questions can be made self-explanatory, then the respondent can be left to complete them unsupervised. This they can do in their own home, forwarding their reply, by post, when they have finished. Hence, the methods that allow the respondent to reply by post are, by definition, the most straightforward. While the most common way of eliciting personal information (in written form) is a postal survey, there is no reason why one can not simply allow respondents to 'tell you in their own words'. This obviously depends on research question lending itself to this mode of reply. Both will be discussed in the following section.

Surveys

The History and Social History of Surveys

Let us begin by looking at what it means to survey (say) a piece of land. Surveying (a piece of land) involves finding a suitable vantage point from which one can view the terrain, and then mapping the geographic attributes of the area. A social survey is not all that different – instead of mapping the geographical attributes of the area, one 'maps' the social attributes of the 'area'. As condescending and simplistic as this may sound (to contemporary minds), the very first social surveys were just that. The very first social surveys were simply street maps, coloured to show the affluence (poverty) of the residents (Bulmer, Bales and Sklar, 1991). Reproduced opposite is a section from Charles Booth's *'Descriptive Map of London Poverty 1889'*, taken from his study of *'Labour and Life of the People in London'* (1891). (It is also the area where the *'Jack the Ripper'* murders took place, which were also about this time.) Booth 'employed' school board visitors to assess the financial status of each household. He then used this information to create a seven-fold classification of poverty/affluence. The poverty/affluence of each household (or area) could then be displayed on a local street map: each level of poverty represented by a different colour. Poverty was one of the major social problems of the day; the affluent classes were aware of this problem, but they lacked the information to deal with it effectively – all they had were a few anecdotal reports. Booth's survey showed them where the poor of London lived (Bulmer, Bales and Sklar, 1991). More importantly, it established a research design, which is still used to this day. The Government's census is still about mapping poverty/affluence: occupation, car ownership, type of property, residents per household, etc. are all indicators of affluence (or poverty), and they can all be classified according to their location. Another example is *'The British Crime Survey'*. This is basically a 'map' of another social problem – crime (for the 1992 edition, see Mayhew, Maung and Mirrlees-Black, 1993).

In contrast (to the surveys commissioned by government organisations), the surveys commissioned by commercial organisations are designed to elicit mass opinion. In order to understand why these organisations conduct such surveys it is perhaps useful to look at how and why modern society has come to value mass opinion. This means looking at what a 'modern society' is. The word 'modern' (as a lifestyle), conjures up images of labour-saving devices, leisure technologies (such as television), the latest fashion, conspicuous consumption, and so on. This kind of *lifestyle* would not be achievable (for most people) without mass production, and mass production is only viable with sufficient demand. But how can a manufacturer know if there is, or will be, sufficient demand for the product? The answer is, of course, by *consumer* research. Let us suppose, however, that it is bad news for the producer (there is not enough demand for the product). The producer can either accept that the venture is not economically viable, or attempt to create more demand for the product. This can be done by advertising the product (in the *mass* media), which (often) involves portraying it as part of a 'modern' lifestyle! (for social commentaries on the sociology of advertising see: Dyer, 1992, *'Advertising as Communication'*; Leiss, Kline and Jhally, 1990, *'Social Communication in Advertising: Persons, Products and Images of Well-Being, Second Edition'*; and Williamson, 1993, *'Decoding Advertisements: Ideology and Meaning in Advertising'*). From this, we get things such as fashion (see, for example,

Acknowledgements to Charles Booth, London Topographical Society and Colchester Colour Imaging.

Wilson, 1987 and also Harman, 1985), fads (Aguirre, Quarantelli and Mendoza, 1988), and body image (see, for example, Featherstone, 1982). Robert Dunn, however, gives a more political interpretation:

> The conscious effort to turn workers into consumers initially was instigated by the crisis of early twentieth century capitalism, in which the need to sell products on a mass scale coincided with the need to control an increasingly dissatisfied and politicized workforce. The ideology of consumption thus originally sought to create a manageable labour force focussed on the desire to purchase consumer goods.... The crucial role which mass media were destined to play in this process was fully revealed by the widespread application of the camera. Photographic visualization gave advertising the capacity not only to present persuasive product images but to project powerful cultural and psychological meanings. Through visual images advertising moved far beyond the mere supplying of 'consumer information' to the selling of desire itself. Advertising soon became less a sales effort for specific products than an expression of lifestyle, self-esteem, and social order.
>
> ***Dunn, 1986:50-51***

What this is describing, is a period in our *social* history known as *Modernity*. Modernity is a mass produced society in which culture is homogenised and consumerism is the *raison d'être* of its members. It is a society preoccupied with being 'modern', where everything and every*one* has its price. Its members sacrifice individuality for the sake of being able to "keep up with the Jones's". Obviously there is more to the Modernity *thesis* than the few ideas sketch here, but this gives some idea of the kind of society, which has given rise to the 'attitude', 'consumer', and 'public opinion' survey. (The *'Modernity Thesis'* the name given to the body of (sociological) literature, which explains how all components of society (from this particular period in our social history) fit together. Peter Berger has written several books on Modernity, and is a good reference for those interested in a broader discussion of the Modernity thesis.) Mass research (the survey) is a part of a mass produced society, which (according to the Modernity Thesis) makes it as much a tool of society as it is of the researcher. The survey method is, therefore, both a characteristic *and* a consequence of Modernity.

Design Validity

An in-depth discussion of question wording and survey design is beyond the scope of this book. (It would require several texts, of this size, to do the subject justice.) There are, however, a number of very good texts on the subject, the most popular of which are: *Asking Questions: A Practical Guide to Questionnaire Design by Sudman and Bradburn* (1989); *Questions and Answers in Attitude Surveys: Experiments on Question Form, Wording, and Context* by Schuman and Presser (1981); and *Questionnaire Construction* by Gillian Courtenay in *Survey Research Practice* by Hoinville and Jowell (eds.)(1983). In addition, academic journals, such as *Public Opinion Quarterly*, regularly publish articles on questionnaire design and methods. There is, however, something to be gained from looking at this subject, in the context of the above discussion(s). As with all of the discussions, in this part of the book, the emphasis will be on rationale. This emphasis on rationale will be extended into the third section, with a (brief) discussion on response rates.

In order to understand the principles of survey design it is necessary to think about what a 'survey' is designed for. Social surveys tend to deal with 'factual' information, which makes the design and purpose of a social survey virtually self-explanatory. This is not the case with an attitude survey. Attitudes and opinions are mental processes, as such, they are not amenable to direct observation. (Remember the distinction made between the researcher's control over the data in this chapter, compared to the designs discussed in the previous chapter.) Therefore, the 'attitudes' and 'opinions', elicited by an attitude survey, are not actually attitudes and opinions, but inferences, made about the same, based on the way in which the respondent has answered certain questions. The first, and most obvious, question this raises is: *'to what extent do these questions (taking into account, any limitations placed on answering them) reflect the attitude or opinion they are supposed to?'* (This is what researchers refer to as **validity** or, more precisely, **internal validity**.) Clearly, this is a very difficult question to answer. If no measure of the attitude/opinion currently exists, then 'validity' is established by an absence of *invalid* questions/items. While it may not always be possible to say whether or not an item (or instrument) is valid, one can usually be more certain about items (or instrument) that are (obviously) not valid (is not tapping in to the target attitude/opinion). (This will be dealt with in the section on design.) This, of course, assumes that the attitude/opinion can be identified with sufficient precision (and distinguished from similar attitudes/opinions) for this to be possible.

The second question is: *'how is the respondent interpreting the question, and what is being asked of them?'* This clearly has two components: the respondent's understanding of the question (this will be dealt with in the section on design), and how *they think* the researcher wants them to answer the question. The second component can be further broken down into: demand characteristics, impression management, self-knowledge, and role occupancy. Again, it is not always possible to know whether a respondent has answered questions: 'as they are', 'as they would like to be seen, 'as they like to see themselves', or 'as they see themselves while occupying a particular (social) role'. The third, and final, question relates to what researchers call **external validity** (it is also an on-going topic in social psychology): *'to what extent does the attitude measure predict a person's behaviour?'* The instrument may be a good measure of the target attitude, but this does not tell us how the attitude relates to the person's behaviour. There are theories and measures which predict this (we will look at them later), but again it comes down to the researcher being clear on the (research) question they are seeking to answer. Before looking at the mechanics of survey design, let us see how these issues could effect the research.

Much of the research into survey design is about how different responses can be elicited from differently worded questions. It obvious that these must be tapping into something different each time. For example, Schuman and Presser (1981) found that 'forbid' and 'not allow' elicit different responses (see also Hippler and Schwarz, 1986). The same is also true of 'abortion' and 'ending a pregnancy' (Schuman and Presser, 1981). This begs the question: *'which of these (terms) is the most valid?'* Clearly the answer depends on the research question. If a researcher wants to know how people feel about terminating the life of an unborn child, then 'abortion' is the expression to use. On the other hand, if a researcher wants to know how people feel about a woman having the freedom to end an unwanted pregnancy, then 'ending a pregnancy' is the expression to use. However, if a researcher wants an opinion, which considers both sides of the issue, then they would use both expressions. Different questions will produce different responses, therefore, subtle differences in the wording of the research question will be translated into slightly different questions on the survey/questionnaire. Arguably, the problem is not with the wording of the questions, but with the researcher being clear on what it is they want to know. A few years ago, one of my students was interested in investigating people's attitude towards (career) women, who wanted to have children but did not want a husband. Although her survey made it quite clear that this was the focus of the investigation, it referred to these women as "single mothers". The responses were less than positive. In this case, the respondents may have found it difficult to ignore the negative connotations associated with the phrase 'single mother'. It is not always possible to know whether a question will be read as it has been written (as the researcher intends it to be read), but it is something that one always needs to consider.

The design of a survey should, therefore, begin by operationally defining the research question. In this case, operationally defining the research question means knowing what one wants from the survey/questionnaire. If the research question is asking for how people would *behave* in a certain situation; asking them how they would *feel* in that situation (and inferring their behaviour from that), may not be a valid interpretation of the research question. Even if a researcher can claim to 'know' how people's feelings are related to their behaviour, this still does not tell us how they would *actually* behave in that situation. (The researcher is still making an inference.) The operationalising of the research question is important (for the validity of the research) because it provides the focus for the next stage in the process - question/item generation. In terms of the questions/items, themselves, the researcher needs to ensure that the mental processes are faithfully reproduced by the responses to the questions/items. Moreover, they need to ensure that the questions/items will be understood, in the same way, by everyone. (These issues will be discussed, in detail, later.) An interesting example of this can be seen in series of studies beginning, in 1939, with Clark and Clark. Their procedure consisted of giving an Afro-American child two dolls (one doll was black/Afro-American, and the other was white/Caucasian), and asking them questions about how they felt towards those two dolls. (It was assumed that the child's own feelings of self-worth would be reflected in their behaviour towards the dolls.)

The questions were as follows:

i) Give me the doll that you want to play with.

ii) Give me the doll that is a nice doll.

iii) Give me the doll that looks bad.

iv) Give me the doll that is a nice colour.

After Clark and Clark, 1947

The researchers noted the percentage of children choosing the black doll after each question (which supposedly indicates a positive self-image). The study has been replicated a number of times since, and the results from three such replications are shown below (together with Clark and Clark's original findings). In addition, the table shows the dates when these studies were *conducted.* (The reference shows the date of publication.)

		Clark & Clark (1939)	***Gregor & McPherson (1964)***	***Asher & Allen (1967)***	***Hraba & Grant (1969)***
Play With	**Black**	32%	49%	23%	**70%**
	White	67%	51%	76%	**30%**
Nice Doll	**Black**	**38%**	**59%**	**30%**	54%
	White	**59%**	**41%**	**69%**	46%
Looks Bad	**Black**	59%	60%	73%	36%
	White	17%	40%	24%	61%
Nice Colour	**Black**	**38%**	**59%**	**29%**	**69%**
	White	**60%**	**41%**	**69%**	**31%**

After Clark and Clark, 1947; Gregor and McPherson, 1966; Asher and Allen, 1969; Hraba and Grant, 1970.

One can see, from the above table, that the percentages differ depending on the question asked. The percentages for the questions relating to the perception of the doll and its colour, differ from the percentages for the question relating to playing with the doll. In the 1939, 1964 and 1967 studies, the percentages for the 'Nice Doll' and 'Nice Colour' questions are almost identical. This suggests that the two questions may be tapping into the same underlying opinion. Unfortunately, this data does not tell us whether it was the same children who were giving the same responses in each case (this is a problem with aggregating responses). The consistency of the percentages suggests that this may be the case. (The probability of the children giving different responses, but those responses balancing each other out across the three investigations, would be a very rare event indeed.) However, without more information this is still just an assumption. The same is true for the 'Play With' and 'Nice Colour' questions, in the 1969 experiment. On the face of it, 'Nice Colour' appears to be a good predictor of 'Play With', but, here again, we can not make that assumption. This, brief discussion, raises several question about the validity of such comparisons. Liking for the doll may not reflect the child's own self-image. In this case, the child may have simply been feeding back, to the researcher, what were the prevailing ideals of attractiveness and likeability, at the time. It is assumed that these studies show an increase in positive self-image among black (Afro-American) children, but could it also show racial segregation. (The percentages for 'Nice Colour' and 'Play With' are almost identical in the 1969 study.) This shows why it is important to be clear about the nature of the relevant (to the research question) constructs, and whether or not they will be evidenced by the design questions/research instrument.

This also raises the issue of how people's feelings relate to their behaviour. A very well known study of this was conducted, by LaPiere, in the early part of the twentieth century. At the time (of the investigation), there was a great deal of prejudice against Chinese people (people of oriental appearance). LaPiere took a Chinese couple to 66 hotels and 184 restaurants, in order to see whether or not they would get served. They were refused only once. LaPiere then wrote to the (same) hotels and restaurants asking them if they would allow a Chinese couple to dine in their restaurant. Ninety one per cent said *no* (reported in Berkowitz, 1986:178). This shows that there is often a difference between how people behave and how they think they will behave. This can affect the (external) validity of a piece of research, particularly if it involves predicting people's behaviour. Social psychologists have devoted a considerable amount of time and resources to researching this issue. One of the more recent theories is known as *'The Theory of Planned Behaviour'* (Ajzen, 1991). *The Theory of Planned Behaviour* attempts to predict a person's *'Intention'* (to carry out the behaviour), based on their 'Attitude' (thoughts and feelings towards the issue), 'Subjective Norms' (what others around you think about the issue) and 'Perceived Behavioural Control' (the ability to control one's behaviour under relevant circumstances). This, of course, assumes that 'Intention' (to carry out the behaviour) is a good predictor of behaviour. The *Theory of Planned Behaviour* is not a solution to the problem of external validity; it is simply a way of showing how complex this issue can be. With this in mind, let us now look at survey design.

Survey Design

Survey questions broadly fall into two types: those relating to the construct of interest, and those relating to the attributes of the respondent. It is usually the case that different groups, or different sectors of the population, will have different opinions about the subject of interest. A researcher may have justification for believing this (in the form of existing research), or they may simply want to know *whether* it is the case. To this end, a researcher may include (in their survey) questions about the respondent's age, sex, marital status, number of dependants, occupation and so on. These are known as **demographics**. Social (government) surveys consist almost entirely of demographic questions (focusing, as they do, on factual information rather than opinions). When it comes to deciding which demographics (demographic questions) to include, there are principally two things to consider. The first is that people have to answer questions about themselves, (almost) every day of their lives (and most of these questions refer to demographics). Imagine how they are going to feel being asked these questions, yet again, on yet another questionnaire. In order not to bore respondents, before they have even started, demographic questions should be confined to what is essential (for the research). If a survey looks like it is going to be too time-consuming, then the respondents are unlikely to want to fill it in. Each question (demographics included) needs to be assessed on a cost benefit basis. In other words, whether the utility of the additional information justifies the inclusion of the question. The 'cost' side, of the equation, refers to whether it is likely to deter people from answering the rest of the questionnaire (or indirectly, may make the survey look long or dull). The 'benefit' side, of the equation, refers to how useful the information is likely to be. The second thing to consider is that some people may be sensitive to certain types of questions, particularly those relating to their age and/or marital status. A researcher can either accept that participants have the right not to answer certain questions, or they can attempt to make them less 'threatening'. If the information is essential to the research, then the researcher really has to make the questions as 'friendly' as possible. For example, *'ticking the box'*, to indicate one's age (group), is much easier than having to (physically) write the number of years, *'in the space provided'*.

Most of the questions *on an attitude survey* will, however, relate to the attitude/opinion/ behaviour of interest. In this case, the structure and wording of the questions is even more important. Questions can be differentiated in terms of the degree to which they structure the participant's responses. In this respect, there are basically two types of questions: **open** and **closed** (Schuman and Presser, 1981). Open questions allow the respondent to reply *in their own words*. Closed questions provide the respondent with a number/range of options/responses from which they must choose 'their' response (or responses). For this reason, closed questions are sometimes referred to as **forced choice** questions. There are advantages and disadvantages to using both types of questions. Open questions have the advantage that they allow respondents to express themselves freely, but this can make analysis difficult. It may be that the responses 'naturally' fall into certain categories. Alternatively, they may have to be **coded** (categorised), in order to be analysed. The exception, is when the question is being used to provide a supplementary anecdote to a previous response, for example: *'Did your parents ever smack you a child? Tell us about one particular incident that you remember.'* The disadvantage, with using open questions, is this need to code the responses (assuming that the data is to be analysed in this way). This can be a time consuming and costly process (but it is essential if the data is to analysed using a computer programme). Open questions do, however, maintain the integrity of the data *better* (at least until it is reduced by the coding process). At this point, it is perhaps worth reminding ourselves that any form of (imposed) categorisation, coding, or classification may later become a way of 'seeing' the data. It will also cause the data to lose some of its integrity.

An open question can also be used to determine a range of responses for a closed question. This involves trying the question out on a small number of respondents, in order to see what kind of responses the question elicits, and whether it is suitable for a larger investigation. (This is know as **piloting**.) These responses are then put into categories, which become the choices for closed question. It may seem like the difference between open and closed questions is simply the point at which the data is coded/categorised, but this assumes that responses elicited will/would be the same in each case, and that the coding/classification accounts for all possible responses. There is a number of reasons why this may not be the case. Firstly, the pilot survey may not have the diversity of responses found in a large-scale survey. The sample is smaller (in the pilot study), therefore, the range of responses is likely to be smaller. When an open question is turned into a closed question, the respondent is forced to choose the option, which *most closely approximates* their own (had they had a free choice). Some of these choices will be closer (to what they *actually* think) than others, and this will vary from person to person. Closed questions are, therefore, 'forced choice' questions in more ways than one. In this respect, the validity of the item/question is defined by the fit of the options to the actual responses. This will obviously depend on how the question is worded and the choice (and wording) of the options supplied. (The *order* of the options can also have an effect on the participant's response: see Schuman and Presser, 1981:Ch 2.) The advantage of using closed questions is that the responses come pre-structured. This makes them easier to analyse, particularly when using a computer.

Another way in which 'forced choice' questions may produce invalid responses is through the introduction of a 'don't know' option. On the one hand, a 'don't know' option ensures that respondents are not forced into giving an opinion they do not (really) have, while on the other, it provides a coward's way out for people who are unwilling to commit themselves. For this reason, some researchers prefer not to have a 'don't know' option, thus *forcing* their respondents to make a choice. This can create the (false) impression that more people care about the issue than is actually the case. If something is not *really* an issue, then resources could perhaps be better to spent elsewhere. For this reason, it is usually advisable to ask respondents if they have an opinion, before going on to ask them to quantify it. (Schuman and Presser, 1981:Ch 4). Another problem with the 'don't know' (or if it is on a scale: middle range) response is that it is difficult to interpret (Schuman and Presser, 1981). A neutral response could mean that the person is: a) ambivalent; b) not sure how they feel; c) genuinely neutral; d) unwilling to commit themselves; e) unwilling to disclose their real feelings; or f) ignorant of the issue. It perhaps goes without saying that the need for interpretation is yet another way in which a genuine response can lose its validity. This problem is not only confined to the question of whether or not the opinion exists, but the question of how strong it is. The following is typical (in terms of format) of the kind of item statement and scale found in most questionnaires. (This is known as a **Semantic Differentiation Scale**, credited to Osgood et. al, 1953, by Courtenay, 1983.)

To what extent do you agree with the following statement?

This is a good book.

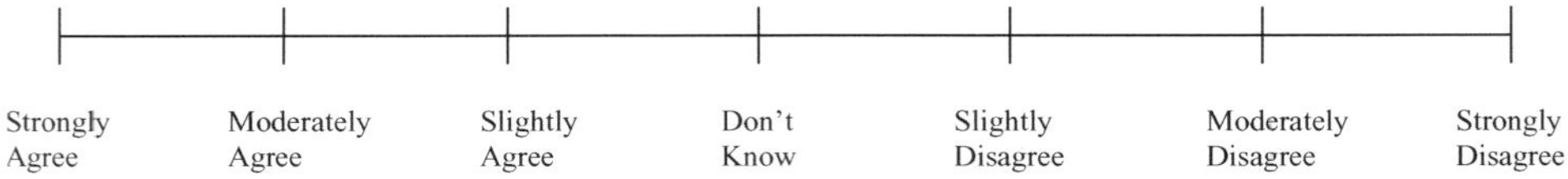

The problem with using this kind of scale is that the increments may be drawn as equal, but they may not be equal *in the mind of the person answering the question*. For example, the difference between 'moderately agree' and 'slightly agree' may not be the same as the difference between 'moderately disagree' and 'strongly disagree', *in the mind of the person answering the question*, and this may differ from person to person. This is what is known as an **Ordinal Scale**: the responses can be ordered, but the increments are not equal (and this has implications for the way in which the data can be analysed – see Taylor, 2007).

Much of what has been discussed, in this section, is not what makes a survey valid, but what makes it *in*valid. When it comes to questions of validity, all a researcher can really say is that they have designed their survey not to be invalid. However, this assumes that 'opinions' and 'attitudes' exist as just that – discrete mental constructs. Could it be that 'attitudes' and 'opinions' do not exist as discrete mental constructs, but are assembled, *ad hoc*, from a mix and match of ideas appended to some vague heuristic? This would explain why it is easier to identify invalid items than valid ones, but it does not explain what it is the research is actually tapping into. Could the 'attitude', or 'opinion', simply be the sum of the answers to this particular set of questions? This aside, people tend to be consistent in their responses to certain types of questions. It is also the case that people can be differentiated on the basis of their responses to certain types of questions, which suggests that the questions are tapping into *something*. This hints at another problem with the survey method, and social research in general: if certain types of people respond to certain questions in a certain way, is it not possible that the people who respond to the survey *per se* may be different to those who do not? It is an even greater concern if the respondents are in the minority. The issues surrounding response rates and respondent characteristics are almost as well researched as the issues surrounding question wording. No discussion of the survey method would be complete without a look at this literature.

Response Rates and Respondent Profiles

The other main issue, occupying the minds of survey designers, is the proportion of respondents who are willing to complete the survey (the response rate). Being able to generalise (the results), from the research sample to the wider population, is as important for survey research as it is for experimental research. However, the impetus is slightly different in each case, which means that the two methods approach the problem somewhat differently. The experimenter has the advantage that his/her method is perceived (by the general public) to be the more 'scientific, and so has authority of 'science' behind him/her. Whereas, the survey researcher has to rely on his/her 'charm' (see below). Appealing to the respondent's altruistic nature is unlikely to get them very far (Brook, 1983). By definition, the experimental method is designed to minimise the effects of interpersonal differences, therefore, any differences, between those people who are willing to participate in the research and those who are not, are unlikely to concern the experimenter. (The purpose of the experimental design being to determine the effect of the independent variable on the dependent variable.) An experimenter is only likely to be concerned about the attributes of individuals who are not willing to participate in the research, when the experiment is in the domain of social psychology (with the emphasis on the social). A survey researcher, on the other hand, needs to ensure that their results are representative of the population as a whole (or the relevant sector). Survey designers have, therefore, become preoccupied with discovering what makes people (not) want to participate in a survey.

Certain features of the survey and/or its administration have been found to facilitate (or inhibit) responses. One of the more obvious factors is length. According to Brook (1983), eight to ten sides (A4 size) is about as much as people will tolerate. It does not seem to make much difference whether the questions are printed on one side, or both sides, of the paper (Childers and Ferrell, 1979). What seems to be important is the overall look of the survey (Brook, 1983). Reading the literature, it seems to be not just the impression created by the survey, but the impression created by the survey *together with* how it is administered. The use of pre-paid envelopes, covering letters (signed by someone in authority), follow-up letters (expressing concern, disappointment and surprise), and thank-you letters, have all been shown to have an effect on response rates (for postal surveys) (Brook, 1983). Inducements, or good will gestures, can also have an effect on response rates: the most common of which is a small gift. These 'give aways' have included: key rings, stamps, lottery tickets, golf balls, photograph holders, tie clasps, packets of instant coffee, letter openers, and pens (review by Nederhof, 1983). As an undergraduate, I participated in a piece of research (conducted by Ian James) in which the incentive was a donation to charity (Comic Relief). (Although this was for an experiment, I would be very surprised if it did not work as a general incentive.) This need to provide incentives shows just how tired, of 'surveys', people have become. One of the major factors must surely be the (now widespread) practice of *marketing* products under the guise of conducting a survey (see *'Suggers, Fruggers and Data-Muggers'* a programme in the 'Horizon' series, broadcast by the BBC on 15th February 1993). This has now become a standing joke:

> ***Narrator (Angus Deayton):***
> In any busy, public place you may be being watched. You may try merging with the crowds, though there's no escape. Some people really fear what happens next.
>
> ***Penny Rigden (Freelance Interviewer)***
> Excuse me, have you got a couple of minutes? Could I ask you a few questions?
>
> ***Transcript of the 'Horizon' broadcast 'Suggers, Fruggers and Data-Muggers', produced and published by the BBC, 1993:3)***

Obviously, some people will be more willing to participate (in surveys) than others. To this end, researchers have found that respondents and non-respondents have the following characteristics:

Respondents tend to be:

i) Favourably disposed towards the survey's aims or involved in the survey subject.

ii) Politically or socially active.

iii) In the higher socio-economic groups.

iv) Receptive to new ideas.

v) Rapid decision makers.

vi) High achievers especially in education.

vii) Used to communicating by post.

Non-Respondents tend to be:

i) Elderly, disengaged or withdrawn.

ii) Live in urban rather than rural or suburban areas.

iii) Feel that they might be judged by the responses they make.

iv) Feel that they are inadequate at supplying the information requested.

After Brook, 1983:137

Although this typology was compiled some years ago, it does point to certain types of people being more likely to participate (in surveys) than others. Consequently, the people who participate in surveys may not be representative of the *population as a whole*. Ironically, it would appear that the kind of people who are most easily persuaded to participate in surveys are the kind of people who are least likely to be persuaded by advertising. Market researchers take note!

The reason why response rates have become such an issue, for survey researchers, is because so many surveys are designed to be 'self-administered'. (Or administered by someone with the minimum of training.) 'Self-administered' simply means that the respondent completes the survey, without any guidance from a trained researcher. This means that the survey has to be self-explanatory, hence the emphasis on piloting, items/question selection, response selection, and administration. Self-administered questionnaires do not give the researcher an opportunity to 'probe' the respondent's answers. This is one of the reasons why researchers often prefer to use interviews. While an interview schedule may appear to be little more than a loosely structured questionnaire, there is one important difference: an interview allows the researcher to pursue lines of enquiry, as and when they arise, whereas a self-administered survey does not. As with the methods discussed in the previous chapter, one can trace an evolution, in design, from closed-question, self-administered questionnaires, to open, unstructured interviews. (Again, this does not mean to say that the methods have (necessarily) evolved from each other, merely that different methods have evolved to suit (slightly) different types of research questions.)

(Researcher Elicited) Personal Correspondence

Personal correspondence (letters, e-mails etc.) can be divided into two types (see Hakim, 1993):

i) Those written to the researcher in response for information.

ii) Those written in the course of one's daily life (as part of a personal relationship, business, dealing with a public or administrative body or in pursuit of a hobby/interest).

Although 'personal correspondence' is often discussed as a single source of data, Hakim has made an important distinction here. Letters written (by respondents) in response to a request for information (from the researcher), are not the same as those written in the course of the respondent's daily life. The former has been written in response to a specific request, and contains only information relevant to that request. Whereas the latter has been written by the respondent, for some other reason, but it happens to record something about that person's life. The method of elicitation is different; the reason for their production is different, the nature of the information is different; and the type of question, to which the information will be addresses, is different. It, therefore, makes sense to discuss these two, quite different, uses of personal correspondence in the context of their individual research designs. Given the typology used here (in this text), this means discussing researcher elicited correspondence, here, under the heading of *'Researcher Elicited Personal Information'*, and personal correspondence, in the next chapter, under the heading of *'Personal Records, Documents and Accounts'*.

In the course of my investigation into workplace celebrations, I wrote to a number of organisations, asking them what contribution workplace celebrations make, to working relationships, in their organisation. My letter read as follows:

> Dear Sir/Madam,
>
> I am conducting independent research into unofficial workplace celebrations. In other words, the type of things done to mark a person's birthday, leaving, marriage, etc. or at Christmas. I would be interested to know what effects you think it would have in your organisation if a decision was made that these would no longer be permitted to continue (for reason of disruption to work, costing the organisation's time, etc...
>
> Yours faithfully,

This yielded a number of responses, the most interesting and informative of which are as follows:

Example 1:
Letter from a senior female officer in a Social Services department

> Dear Mark,
>
> Thank you for your letter which I received on 21 February 1994.
>
> In my organisation, which is a Local Government Social Services department, workplace celebrations are usually limited to "Leaving Dos". They vary with the length of service and, I think, the psychological status of the person concerned. By that I do not only mean place in the hierarchy, though I think that mostly they match. They are planned, and can include some alcohol, food etc.
>
> Occasionally opening new services etc are celebrated, but by default they are less personal.
>
> If they were stopped, the less formal, more personal celebrations would happen anyway, but elsewhere. My feeling is that they would confirm work/change weary staff feelings that the organisation does not care about them as people. It's a strange job, people's motivation is to doing a good job for their clients, but they

want the organisation to do a good job for them, and an organisation cannot do all of what is expected. Official celebrations would be missed if denied, and as I said above, would probably fit with the rather cynical view of many staff. However, this would not fit the powerful motivating sources that drive social workers. It could contribute to the ever present fear of burn-out but only on a modest scale. other factors are more powerful. However, as surviving in this job is so difficult, I would not stop such practice as there is.

I hope this is of some use.

Yours sincerely

Example 2:
Letter from a male manager in a Job Centre

Dear Mr Taylor

Thank you for your recent letter about workplace celebrations.

We are a relatively small unit working very closely together as a team and, as such, relationships between staff are bound to develop. Birthdays, weddings promotions, retirements etc are often seen as causes of celebration and I have little or no general objection to this as they provide a little light relief for my staff who constantly work under considerable pressure. The real art, for me as a manager, is to minimise the impact these celebrations have on the work of the office and this I do by encouraging celebrations to be held during the lunchtime or towards the end of the day and my staff seem to be quite happy with this arrangement.

I would be very reluctant to see any move to ban or control these celebrations. They are very much a part of the normal pattern and have decided benefits in fostering good working as well as personal relationships, they provide an opportunity for staff to let off steam, they allow management and workers to get together on a social level away from the authority of the workplace.

I am a great believer in the old adage 'a happy team is an efficient team' (and it works harder). Staff need to be treated with respect, need to be treated as human beings and they deserve some consideration. Let's keep some informality in the workplace, keep a human face in the office.

Yours sincerely

Example 3:
Letter from a female secretary in a "fairly large company"

Dear Mark

In response to your recent letter requesting information on unofficial workplace celebrations, I have the following comments to make.

Firstly, my position within the company is Secretary.

These celebrations refer to our small section within a fairly large company.

Birthdays
Each department gives a card to the person celebrating their birthday, this person then provides a cake for each employee. Some employees like to go to the pub with their particular friends for a drink at lunchtime but this is not obligatory.

There has only been one deviation from the above, when I had my fortieth. My desk was decorated with balloons and streamers. I was given flowers by two departments and my husband was allowed to organise a Kissagram (sic) during working hours.

Leaving
A leaving card is signed by all those giving to a collection. On the appointed day a presentation of the card and present is made about 15 minutes before breaking for lunch. The presentation is normally carried out by the most senior member of staff on the premises at that time. Everybody is invited to the pub.

Marriage
As above.

Christmas
Up to two years ago we were a company on our own and each Christmas we finished work at lunchtime on our last day and were treated to Christmas dinner and wine at a local hotel, being allowed to stay as long as we wanted to socialise. It was great fun.

As we are now part of a large company and they could not afford to do it for everyone, it had to stop. Within our section we do normally organise an evening meal but this has to be done by one of the staff and we have to pay for ourselves. On the day we finish for Christmas, we go to the pub because we have to go back to work and smelling of "booze" at work is frowned upon, the celebrations are very low key. We finish work at the usual time with no concessions.

If we were told to stop any of the above activities, I do not think it would cause to (sic) many problems, we would just have to make sure that they were carried out during our own time, ie coffee and tea breaks and lunchtimes and Christmas does not interfere with company time or cost them anything anyway.

I hope this information is what you require and is of some use to you. I wish you luck with your research.

Yours sincerely

These letters tell us how different people, of different status, see this type of event (or saw it at the time of publication). They also show a difference in the way in which the three authors interpreted the task of reporting it (to me). The two local government managers used it as an opportunity to mention how hard their staff work, and how fragile the morale of their staff is. This tells us *something* about these people, their role and position in their organisation(s), and the role of their organisation within society. As managers, not only is it their job to 'keep the troops happy' (as they rightly point out), but to portray their department as an efficient, economical, and productive unit, not just to their own senior managers but to the society their department serves, and this is precisely what they are doing here! For them, as managers, 'morale' is an issue, and they see workplace celebrations as an opportunity for their staff to take a break from the work and to get to know each other (cementing working relationships). They believe that their staff see workplace celebrations as a way of 'letting off steam', and getting to know one another. As such, they are not telling us about how their staff see such celebrations: they are telling us about how *they think* their staff see such celebrations. They are obviously aware that banning such celebrations would not be a popular decision, but they would not be able to stop them.

The third author is a secretary (one of the 'workforce') and , as such, workplace celebrations are of personal relevance to her. This is clearly reflected in her correspondence. Her account describes *the norm* for celebrating birthdays, in her organisation, as well as her own fortieth birthday celebration, which was an exception. She also tells us about how Christmas celebrations have changed, as a result of changes in the structure of the organisation and company policy. One can tell how she feels about this. Her account includes how she thinks others, in her organisation, see such celebrations. In this case, the 'others' are her new managers, and her perception is that they see it as a waste of money. Although, these accounts provide us with three very different views of workplace celebrations, it is perhaps more accurate to say that they provide us with three accounts of workplace celebrations, written in the context of the author's perception of their organisation. The way in which these authors write about the subject tells us something about how *they* view it, their position in the organisation, and what kind of people they are. This suggests that the validity of such accounts may be open to question. The fact that the question of validity has been raised, naturally, begs the question: '*what makes one suspect that the account may not be valid?'* The answer to this question, then becomes the context in which the information is interpreted, and this, in turn, becomes part of the analysis itself (as demonstrated here). Methodological issues aside, this research shows that if the topic can be reasonably well defined, then there is no reason why the respondents can not simply put the information in a letter (or even an e-mail).

(Researcher Elicited) Diaries and Diary-Interviews

The idea of *researcher elicited* correspondence can be extended to *researcher elicited* diaries. While the above procedure gives respondents a relatively free hand, it can afford to do so because the topic is quite specific. The letter sent to respondents did, however, contain a certain amount of guidance: "I would be interested to know what effects you think it would have in your organisation if a decision was made that these would no longer be permitted to continue." What these examples do not show, however, is the variation in the quality of the reports – some were literally a sentence or two. Simply asking respondents to send a letter, or to keep a diary, does not mean that the depth, or format, is going to be the same. When asking respondents to keep a diary it is, therefore, necessary to specify the kind of information required, and even the format. A good example of this can be found in the work of Maas and Kuypers. Maas and Kuypers used a highly prescribed form of the researcher-elicited diary in order to investigate people's adjustment to old age. The design was standardised to the point where the respondents were provided with notebooks, with the pages (a page for each day) divided into "morning", "afternoon" and "evening". In addition, the respondents received the following (standardised) instructions:

> We would like you to keep a daily diary to help us get some idea of how you spend your time during a typical week. We are especially interested in the kinds of things you do, when you do them, for how long, and whether you do them with other people. As you write your diary, be sure to include the time of day when you get up, have your meals, go out of the house, and any other major activity. Also be sure to include whom you met, what you did with them. You may also want to include some of the thoughts and feelings that you had during the day. At the end of the week, look over the diary to see if you have described a pretty typical seven days. Make any comment you want about what you have written…
>
> ***Maas and Kuypers, 1974:218***

This is just one of the ways way in which (researcher-elicited) diaries have been used by researchers. Diaries have also been used as the basis for interviews (Burgess, 1993; Zimmerman and Wieder, 1982). In this case (the diary-diary-interview method), the diary is seen not a source of information *per se*, but as a way of exploring the (research) subject (Zimmerman and Wieder, 1982:127). The following example shows how this can be used to good effect:

Example 4:
Diary and Diary-Interview extract from Zimmerman and Wieder (1982):

> (O)ne of our (Zimmerman and Wieder) diarists wrote in her diary:
>
> Talked to S.H. about the bathing suit which she is going to batik for me. She "got it for me wholesale" from [local store], and I will probably trade something for her doing the suit.
>
> The researcher first read this section of the diary to the diarist, hoping that she

would 'spontaneously' elaborate. But the diarist replied, "Mm-mm," requiring the researcher to be more direct in her questioning:

> ***Researcher:*** Could you tell me more about that – is that a typical kind of thing that you, like, sort of exchange goods?
> ***Diarist:*** Yeah, yeah, it's very typical. Ah, I live, around, and it always seems that I've known a lot of people who make things and do things, and so we just exchange one skill for another. Somebody made me a pair of earrings once, and I dyed some cloth for them, with natural dyes, and, ah, ah, especially in where I live now, you know…there are lots of people around that exchange skills, and nobody seems to have a lot of money, so we just, ah, L. gave me a drawing once, and I gave her some homespun wool so she could do a weaving, and then she did me – she did a weaving for me, and I crocheted her a hat and gave her some homespun to do some more weaving.
>
> The researcher would then listen to the informant's response and then probe again for more detail if the developing description was relevant to the research concern.
>
> ***Zimmerman and Wieder, 1982:127-8***

Zimmerman and Wieder (1982) actually intended the method (the diary-diary-interview) as a pilot for ethnographic (or participant observation) research. (The main investigation being designed around the information provided by the diaries, which would be kept by 'assistants/informants' in the field.) It is an efficient way of eliciting a range of perspectives, some of which would not normally be available to the researcher. Used in this way, the diary and diary-interview provides a focus for prospective research.

The diary-diary-interview method has also been used as method in its own right. For his research into secondary (junior high) school teaching, Burgess (1993) used the (researcher elicited) diary as the main source of information, supplementing it with interview questions where necessary. The following shows how this worked in practice:

Example 5:
Diary and Diary-Interview extract from Burgess:

> ***Diary Entry:*** The class worked quite well generally and most people wrote about half a page – then bought the book to me and I wrote an extra question. For some reason this produces a better result…in all, quite a good lesson. Some laughing and gossip from the girls but the boys wrote or talked very quietly…
>
> ***Burgess:*** For the 6th March in an English lesson where quite a few pupils did some writing you commented… "the class worked quite well generally"….I wonder if you could expand for example the phrase "quite well"…?
>
> ***Respondent:*** When I think that a lesson has gone well, I think what I mean is that if I have asked them to write they have done some writing, which certainly when

I asked them to write not all of them always did...

After Burgess, 1993:132 and 133

One thing which stands out about this particular extract is way in which the diarist draws attention to differences in the boys' and girls' behaviour (see Burgess, 1993:130-132). Sex differences, in academic achievement, is a very high profile issue. Perhaps Burgess was being diplomatic here, but he could have used the diary-interview (part) to take this explore this issue in light of the comments made by the diarist.

The diary-diary-interview is useful way of exploring the 'field', but as a 'stand-alone' method, the inconsistencies in its approach make it difficult to apply. The (researcher elicited) diary gives respondents an opportunity to set down what they see as important in their day/week. The interview allows the researcher to focus on what they see as important. In this respect, the two halves of the procedure appear somewhat inconsistent; either the researcher is conducting an analysis of the respondent's account, or they are seeking to elicit a discourse in response to a particular research question. Surely, they can not be doing both? This inconsistency is exemplified in the line: "The researcher would ... then probe again for more detail *if the developing description was relevant to the research concern*" (Zimmerman and Wieder, 1982:127-8, emphasis added). It seems to suggests that the respondent has a (relatively) free hand writing the diary (see Zimmerman and Wieder, 1982:125; Burgess, 1993:130, for examples of the kind of guidance given to diarists), but the researcher will only use that which is relevant to the research question. With this in mind, let us now look at the interview as a method in its own right.

Researcher Elicited Personal - Verbal

There is basically only one method which fits this description and that is an interview. While the method can be adapted for use in a group discussion situation (a focus group), the principles are basically the same. The interviewer's role, in each case, is to facilitate a discourse, which contains the information needed to answer the research question. A good interview is not simply a question and answer session, it is a discussion in which the interviewer uses a few, well-chosen, questions to guide the 'conversation'. Again, the 'written'/'verbal' distinction can be seen to reflect the nature of the response.

Interviews

It is sometimes difficult to draw a line between a verbally-administered survey, with a high proportion of open questions, and an interview. In many respects, the interview is an extension of the survey method, and the points made about design validity (in the above section) apply here as well. As with a survey, the research should begin by articulating the research question *in such a way that it can be answered by the participant's responses to certain questions.* Interviews tend to be more selective than survey's, and this selectivity is reflected in the need to identify respondents with the necessary information/knowledge. In the case of a survey, the researcher has reason to believe that certain questions will elicit certain information. (When it comes to demographics, these are quite obvious.) An interview is based on the same principle, but allows a researcher the flexibility to pursue lines of enquiry, not previously considered. The main difference between survey questions and interview questions is that survey questions are (mostly) 'forced choice' questions, whereas interview questions are designed to elicit a discourse (which should come mostly from the participant). Interview questions should, therefore, allow the participant to speak freely; they should not pre-suppose a particular answer, nor should they be 'threatening'. This applies as much to their delivery as to their content. A good interview is evidenced by a flowing natural discourse in which the interviewee does most of the talking: in response to a few carefully chosen questions from the interviewer. I have found (from experience) that if an interviewer can resist the temptation to fill uncomfortable pauses (usually with another question), the interviewee will usually find something else to say. (Watch a celebrity interview, on television, and see how much time the 'guest' spends talking, compared to how much time the interviewer spends talking.) As with an observational study, it is important to check that all recording devices are set up and working correctly. It is also a good idea to take notes (just in case something goes wrong with the equipment). If the interview is to be transcribed, then allow **Ten** to **Fifteen** hours of transcribing for every hour of taped interview. This *is* how long it takes!!

It is difficult to demonstrate what a 'good' interview looks like, using this medium (a written text). It is, however, possible to offer some discussion based on published transcripts of the same. The following examples are taken from a book entitled *'In the Field'*, by Robert Burgess (1993). The first example comes from an interview conducted by Roy Nash; and the second comes from an interview conducted by Burgess himself. They are both interviews of school children, although in rather different circumstances(as will be discussed). Burgess uses them as illustrations of 'good' and 'bad' interviewing techniques (the good one being his own), but they show a great deal more than this. These examples show how the topic of conversation, the number of interviewees, and the researcher's approach, can all have an effect on the information elicited.

Example 6:
Nash interviewing 'Bruce' (1973)

Int: You know when you filled in that questionnaire for me, what did you say you wanted to do when you left school?
Bruce: Nothing.
Int: Nothing? No, you didn't. I meant to ask you about that. I said last night, you know, I said, 'I'm going to screw his ears off in the morning'. What – you never put your friends down either, did you?
Bruce: No I've not got any friends.
Int: Why not?
Bruce: Because I haven't none.
Int: Got no friends at all?
Bruce: No?
Int: But you must have someone to play with?
Bruce: No.
Int: Well, you sometimes sit by Matthew, don't you?
Bruce: But I never see him. He always goes away and plays with somebody else.
Int: Does he?

Bruce: Yes.
Int: What do you do at playtime then, for goodness sake?
Bruce: Just sit there against the wall.
Int: Do you?
Bruce: Yes.
Int: Haven't you got any friends?
Bruce: No.
Int: Don't you want any?
Bruce: Not really.
Int: Did you have any friends at School E?
Bruce: Yes, I had a lot of friends at E.
Int: You had a lot of friends at E. Well, where are they now, then?
Bruce: I just sort of fell out with them.

After Nash, 1973:100-101, cited in Burgess: 1993:119-120

Example 7:
Burgess interviewing 'John Slattery' and 'David Owen' (1974)

Int: Now we've got John Slattery and David Owen.
John: That's right.
Int:And John you're leaving at Easter and David you're leaving in the Summer?
John: That's right.
David: Unless I get a job then I can go on a three day week.
Int: Oh, I see.
John: Peter Vincent, you know Peter Vincent don't you, he's on a three day week until he leaves school.
Int: How's he getting on?
John: Well, he's getting on all right, he's getting just under £10 a week [in 1974].
David: £10?
John: Yes.
Int: That sounds good.
John: I tried at the place but I never got a job. I've got a job now for when I leave school.
Int: What are you doing?
John: In this car factory place making seats for the cars and the tractors and all this rubbish.
Int: Are you pleased about that?
John: Not really, no. It's only a job see 'cos I'm trying to get some money to go on my holidays. I'm getting this job see and me brother-in-law's getting me a form to fill in for his firm. If I can get in there I'll pack this job in.
Int: But anyway you have got something to go to as soon as you leave.
David: I tried for four jobs last week, I went to three interviews, to a toolmaking place.

John: Good job toolmaking.
David: Yes I know. They all said the same thing you know, they can't insure me until I'm 16.
Int: Yes.
David: You know the toolmaker said that if I wanted to go back when I'm 16 I'd be starting off on £18 a week and I went down another place and they said they can't insure me until I'm 16, they said the same thing. I'd get about £10 or £12. When I'm going to start work the wages are secure on £10 a week. Piece work you see so it depends on how much you do, so you can earn up to £1 an hour.
Int: That's okay.
John: As long as you work hard.
David: But I am guaranteed £10 a week.
Int: That's all right. Well that's about when you have left school. Let's go right back to the schools that you went to first.
John: This school?
Int: Well before you got to this school. Schools that you went to before you came here.

After Burgess: 1993:109-110

The Burgess interview is a good illustration of how an interview should progress. It starts off rather like a question and answer session but it is not long before the boys are talking almost continuously with only the occasional interjections from the interviewer. In the Nash interview, the interviewer does most of the talking. The boy was obviously not in a very talkative mood but the interviewer could have handled this. Burgess used these interviews to illustrate how certain types of questions can better facilitate, or inhibit the discourse of the interviewee, and on the face of it, this is what they show. However, there are other differences (between the two interviews) which may contribute to the efficacy (or otherwise) of the questions. Nash was interviewing a boy who had just changed schools, and had left many of his friends behind. The boy was clearly finding it difficult to make friends, at his new school, and was obviously very unhappy about the situation. Burgess was interviewing two boys who were on the point of leaving school, each with the prospect of a good job. They were clearly in a buoyant mood. The other main difference is that Nash was interviewing one boy, whereas Burgess was interviewing two. In the Burgess interview, the two boys often fed off each other (one boy often responded to the comments of the other – see above), whereas in the Nash interview, the interviewer had to keep the conversation going, on his own. Arguably, Burgess had an easier time of it, anyway.

Aside from these differences, there are a number of ways in which Nash's approach, and style of questioning, could have been improved. His first, and most obvious, failing is his lack of patience. Nash is understandably frustrated by the boy's reluctance to talk, but he could have shown a little more understanding in this situation. Moreover, his style of questioning made the boy feel even more of an outcast, causing him to become even more withdrawn. This can clearly be seen in Nash's questions, and the boy's answers, towards the latter part of the interview (reproduced again below):

Int: What do you do at playtime then, for goodness sake?

Bruce: Just sit there against the wall.
Int: Do you?
Bruce: Yes.
Int: Haven't you got any friends?
Bruce: No.
Int: Don't you want any?
Bruce: Not really.
Int: Did you have any friends at School E?
Bruce: Yes, I had a lot of friends at E.
Int: You had a lot of friends at E. Well, where are they now, then?
Bruce: I just sort of fell out with them.

Nash should have encouraged the boy to talk/open up more. He could have asked the boy what he saw as the difference between the two schools, or how his circumstances have changed. This is the kind of question which it is difficult not to give an elaborate answer to; and is the kind of questions which might have encouraged the boy to have opened up. If Nash was interested in finding out what the boy wanted to do when he left school (as he appears to be), a better line of questioning may have been to have asked the boy about his (after school) interests. This may have actually been a more valid line of questioning than the one he was using. In many ways, this brings us back to the point about operationally defining the research question, and designing the research to answer the research question. In this case, not only did Nash's style of questioning leave a lot to be desired, but his line of questioning was unlikely to answer his (presumed) research question. The Burgess interview was a lot less 'trying' for both parties. More importantly, it demonstrates the advantages of interviewing two, or more, people together. This can be taken one step further, and developed into a group (or a **Focus Group**) interview design. Although the group interview has its advantages, there are a number of problems specific to group interview situations, and focus groups. For this reason, it is worth discussing the method as a separate design.

Focus Groups

In many respects, focus groups (as a method) are simply an elaboration of the interview method. Both have, essentially, the same approach to answering the research question. Eventhough the kind of (research) questions, to which they are applied, are quite different. In this respect, focus groups tend to be used for customer satisfaction research, whereas interviews tend to be used for more social research. The two methods may look different, because the kind of (research) questions to which they are applied look different. However (as they are conceptualised in this typology), they both have the same approach to answering the research question; they are both designed to promote a discourse (group discussion), which will answer the research question. The difference (comparing a focus group to an interview) being that instead of one individual contributing to the discourse, there are a number of individuals contributing to the discourse. As with an interview, it is important to identify those respondents with a knowledge of the subject. In the case of a focus group, that knowledge constitutes a range of opinions on the subject. If the research question (discussion topic) directly relates to people's employment, or community, then there is no reason to believe that people will not be willing to participate in a focus group. (The difficulties are more likely to occur when it comes to arranging a time and location, which is convenient to everyone.) A good example of this can be seen in the group interviews conducted by Ann Lewis. Ann Lewis was interested in finding out how children, in a local primary/grade school, felt about having children with Severe Learning Difficulties (SLD) integrated into their school. (Strictly speaking, she wanted to investigate the utility of group interviews with children this young.) The following extract shows not only the method being used to good effect, but the kind of research questions for which it is most appropriate.

Example 8:
Lewis talking to a group of children about integration (1992)

> Int: Would you like to play with those [SLD] children in the playground here [at mainstream school]?
> *Greg: It would be a bit funny really.*
> *Kathy: Yeh, cos we wouldn't know what to do really. They've got a climbing frame [at the SLD school].*
> *Greg: Cos like they're bigger, but they can't exactly run really quickly and things like that...It's good at their playground because they've got things to climb on so they've got lots to do.*
> *Janine: We've got the infants' climbing frame but we're not allowed on that.*
> *Kathy: We're not allowed, (and) its got concrete underneath, so if they fall off they won't hurt themselves.*
> *Michael: There [under climbing frame, at SLD school] they've got this kind of [gesturing]*
> *Greg: mat*
> *Michael: kind of mat stuff that they have at TG park which is soft.*
>
> Int: Do you think they [SLD children] will always be like that?
> *Sara: (to Cora) I don't think Peter will stop kissing you!*
> *Hannah: He's always kissing me and Cora and other people as well...Some people are afraid to say 'No stop it', they just let it, they don't like, they're afraid to say 'no stop it'.*
> *Cora: I don't like it sometimes.*
> *Emma: It's a bit embarrassing.*
> *Hannah: Yeh, it's embarrassing, I've had loads of people saying 'Ooooh! and everything like that.*
>
> Int:Would you like to have those [SLD] children in your [mainstream] class?
> *Jeanette: I think it would be good for them, but it might stop us doing more things...*
> *Sara:We might start shouting at them.*
> *Hannah: We might scare them a lot...and they might be worried a little bit...and afraid we might hit them again.*

After Lewis, 1992:415

The children were asked how they felt about something which might effect their daily lives (the integration of children with severe learning difficulties (SLD) in to their school), and what the researcher got was a very informative discussion on the subject. It brought out issues of sexual harassment, prejudice (resulting from the SLD children's lack of ability) and uncertainty (not knowing how to behave towards the children with SLD). While the Burgess interview demonstrates how a single co-interviewee can help to maintain a discussion, the Lewis interview demonstrates how *several* peers provide strength and encouragement. As Lewis herself points out, the girls (in the second extract) probably would not have spoke out about 'Peter's' behaviour if they had not been in a group. Once one child felt safe enough to mention the subject, it gave the others courage to do the same. The children clearly felt that they could safely talk about the subject in the group. This is one of the reasons why researchers should ensure that the focus group is a safe environment for people to express themselves. While this example shows how focus groups are an effective way to facilitate a discussion/discourse, the crucial factor would appear to be the (personal) relevance of the topic, rather than the age of the participants.

These extracts also bring out yet another consideration with group interviews: the composition of the group. In this case, the gender of the participants. It is difficult to tell from these extracts, but some groups seem to consist of participants from both sexes, whereas others seem to consist of participants from just one sex. The significance of this may not be immediately apparent, but there is a large body of literature which suggests that men and women use language differently. This difference exists as a general trait, as well as a manifest difference in same and mixed sex conversations. (Jennifer Coates' *'Women, Men and Language: A Sociolinguistic Account of Sex Differences in Language'* (1987) is still one of the best reviews of the subject.) It is not obvious from this example, but "Women tend to organise their talk *co-operatively*, while men tend to organise their talk competitively" (Coates, 1987:154, emphasis in the original). It is, however, possible to see that the girls support one another, which is typical of an all-female discussion (Coates, 1987: Ch9). Lewis, herself, overlooks the possibility that being in (what appears to be) an all-female group, may have also contributed to the girls' willingness to talk openly (about their experiences with 'Peter'). Whether they would have been as forthcoming with a boy in the group, we will never know. Member of the opposite sex can often have an inhibiting effect on the discussion of certain topics.

We have already seen why it is important for participants not to feel inhibited by the presence, or conduct, of anyone else in the group. An opposite sex participant is just one example of the kind of individual who may inhibited a group discussion; others are 'experts', 'hostiles' and 'friends' (Stewart and Shanidasani, 1990). If the group includes an 'expert' (on the subject), then either that person will impose their expertise on the group, or others will simply bow to their superior knowledge. The same can happen when the group includes a participant (or participants) with a particularly strong opinion on the subject. Similarly, while friends may make each other feel more comfortable, it is also possible that they may simply agree with each other. Any kind of dominant influence *can* cause problems, but it is the researcher's (or the person running the group) responsibility to see that it does not. It is the researcher's responsibility to ensure that the discussion is an enjoyable and productive experience for all concerned. In order for this to happen, participants must that they can make a contribution, and that that contribution is valued. If the participants do not feel able to do this, then (at the very least) the discourse will not answer the (research) question, as fully as it might. Again, this can be seen as an issue of validity. It is important for a researcher/group leader to be able to manage a group discussion, but the *list* of personality traits (for a good moderator) suggested by Stewart and Shanidasani (1990; citing Langer, 1978) is perhaps a touch on the idealistic side. While personality is important, group management is a skill which can be learned and improved upon with practice. The optimum number for a group (discussion) is about five, beyond this, things can start to get out of hand. As with an interview, the less the researcher has to intervene, the better.

Conclusion

It is tempting to think that, since all of the above methods are classified under the heading of *'Researcher Elicited Personal Information'*, they are all describing the same process. This title actually describes the origin of the data, the process/method depends on the nature of the research question. Although these methods are all ways in which the researcher can elicit personal information, the differences between them are a reflection of differences in the researcher's role. With (postal) surveys, the researcher's role is essentially one of instrument designer; whereas with group interviews/focus groups, the researcher's role is one of group manager. These methods/procedures can also be seen to evidence of the researcher's lack of control over the data. In this case, lack of control over the data means lack of control over those in possession of the data. This is reflected in the ways in which the respondents/interviewees/participants are *encouraged* (the researcher has no more influence than that) to be forthcoming with the information. In the case of a postal survey, this means giving them an inducement to complete the questionnaire, whereas in the case of a group interview, it means controlling dominant individuals. These can be contrasted to the methods, discussed in the previous chapter, where the researcher only needs to worry about their (own) ability to collect, and convey the information. This brings us on to the subject of the next chapter: *'Personal Accounts'*. In this case, the person making the observation is not a trained observer, nor are they motivated to be impartial. It is, therefore, important to be able to look at personal accounts in terms of who wrote them, why, and what they wanted to communicate, and we will see how this is done in the next chapter.

4: Research Methods and Sources – Personal Records, Documents and Accounts

Introduction

This chapter will look at how information contained in personal documents, records and accounts can be used to answer certain research questions. All the methods, discussed so far, have used data originated, or elicited, by the researcher. With such methods, the data *itself* constitutes the answer to the research question. The existence of the data is explained by the research question, and the research itself. However, from this point onwards, the circumstances surrounding the production of the data become the primary focus. While this is often portrayed as an issue of validity, (as we saw in the section on researcher elicited correspondence) the reasons for doubting the validity of the data are, themselves, a component of the analysis, if not (research) questions in themselves. Therefore, the type of (research) questions, to which these methods are best suited, are those relating to the purpose of the data and the perspective of the author, rather than the information itself. The data can, however, be used to answer a research question, when there is nothing else (better) available. In this case, some data is better than none. Even so, it is important to look critically at the data.

The influence of Positivism has created an emphasis on validity, which, in turn, has created a prejudice against data not gather by the (a) researcher. However, this prejudice has as much to do with social expectations, as it has to do with validity. The popular 'image', of a researcher, is of someone who makes new discoveries. The words 'new' and 'discovery' are not usually associated with something that already exists. Researchers who use existing documents are, therefore, perceived to be shirking their responsibilities, somewhat. This 'prejudice' is not just the 'fault' of Positivism; it is endemic to western culture. For example, performing a song written by someone else ('covering') is perceived to be less difficult than actually writing a song oneself. The ability to perform someone else's song may require just as much skill and ability, but this (type of) skill and ability is not appreciated because our culture values things such as originality, invention, and creativity (and academic research is not immune from this). Given the prevailing (Positivist) (mis)conception about knowledge and the knowledge creation process, it is understandable why a piece of research is perceived to be superior when it uses data gathered by the researcher. (The researcher has gone out and found 'new' data, rather than simply using data from rather dubious sources.) The skills required to analyse an existing document are no less than those required to analyse original data. In many respects, being able to 'deduce' the reasons for the document's existence, the perspective of the author, and the effect both of these have on the information contained in the document, takes just as much skill as it does to analyse 'first hand' data. The task can be just as much of a challenge and just as interesting. With this in mind, let us now look at how these sources can be used and the kind of research questions to which they can be applied.

Personal Written	**Personal Records and Documents, Photos, CVs, Diaries, and (Auto)Biographies**
Personal Verbal	**Oral Histories**

Personal - Written

This section will discuss a range of 'personal' documents. As we will see, the reason for a document's existence is a good indicator of the kind of information it is likely to contain. To this end, 'personal' documents can be categorised into three types: those which *incidentally* record the person's life (these can be further divided into: *administrative* and *personal*), and those which intentionally record the person's life. 'Administrative documents/records' include things such as: applications (for jobs, licences, credit cards, etc.), records/files (police, medical, financial, employment, tax etc.), bills, receipts, and forms of personal identification. 'Personal documents, which *incidentally* record the person's life', include things such as: personal correspondence, e-mails, and evidence of the person's likes and dislikes (clothes, music, hobbies etc.). 'Personal documents, which *intentionally* record the person's life', include things such as: photographs, home videos/movies, diaries, CVs, biographies and autobiographies. Although biographies and autobiographies have been included here, under the heading of *'Personal Accounts'*, most biographies and autobiographies are published. The distinction between the documents discussed in this chapter and those discussed in the next (on published information), is one of personal and public (information), rather than unpublished and published (documents) *per se*. (Journalists often defend themselves by arguing that they are acting 'in the public interest', even when they are reporting on people's private lives.) In terms of validity, it perhaps goes without saying that the documents, which one intends to be a record of one's life, are the one's most likely (out of the three types) to contain an element of impression management. One might also argue that, depending on the type of document and the respondent's motivation to 'bend the truth', certain types of administrative records/ documents are also prone to attempts at 'impression management'.

Administrative Documents/Records

In this context, *'Administrative Documents/Records'* are documents containing information supplied by and/or about individuals, used in the course of an organisation's daily 'business'. This can be contrasted with documents containing information, supplied to the organisation for the purposes of maintaining an archive or a record of the information. (The latter will be discussed in the next chapter – on 'published' sources.) Clearly, this information is 'personal', eventhough it may have been elicited by a public organisation. These organisations can be broadly divided into private sector (commercial organisations) and public sector (the courts, medical profession, local government, educational institutions etc.) establishments. This difference also reflects the degree to which 'the public' (including researchers) is allowed to access the information – the public sector being the *more* open of the two.

The information held by commercial organisations is likely to be that used in the course of their business. It is highly unlikely, therefore, that these organisation will be willing to disclose anything which may be of use to competitors. In addition, the organisation may have given its assurance that the information will only be seen by their employees, particularly if the information is being held on computer. If a commercial organisation is willing to allow a researcher to have access to such information, then (by definition) the information is unlikely to be all that valuable (or informative, from a research point of view). The information held by most commercial organisations is likely to be demographic, and probably not that complete. Interpreting the information may also depend on knowing the history of the document/record. This confounds the problem of interpretation even further:

> Some understanding of the nature and original purpose of any set of records is important not only with reference to the quality, consistency and completeness of the data they provide, but also for the interpretations that can be placed on the results – just as it is crucial to the reading and interpretation of any documentary evidence to know what class of document it belongs to...Obtaining information on how the records were compiled and updated usually requires consultations with the people responsible for the record.
>
> ***Hakim, 1992:42***

The information held in the public sector is usually a lot more accessible (and informative). The most accessible (and widely used) of which is, perhaps, that relating to our legal system: court records, parliamentary records, commissioned reports etc. While (technically) these records may have been produced in the course of 'business' (the 'business' of our legal system), they are often published (as in the case of parliamentary debates and commissioned reports) and/or available for public scrutiny (as in the case of court proceedings). (The legal system has *to be seen* to perform according to established legal doctrines.) This makes legal records among the easiest to locate and study. Consequently, a large number of investigations have used legal records/documents. One of the more unusual of these was an investigation by Neugebauer (1979). Neugebauer was interested in people's beliefs about the origin and causes of mental disorders in the middle ages. Our ability to reconstruct the past is based not only on what survives, but what survives from what was recorded at the time. In the middle ages, the only people who could read and write were the clergy and the aristocracy. Consequently, these are the only views (opinions) recorded for posterity. This means that our view of medieval life is based on those held by the church and the ruling classes (of the time), which gives us the impression that (in the middle ages) mental illness was seen as a punishment from God (for immoral behaviour). The stereotype of the ignorant, medieval peasant, attributing misfortune to witchcraft, demonic possession and supernatural forces, is the one which prevails, not only in our popular culture but in most people's ('modern', 'civilised') minds. Only in the last few decades have historians have begun to question the accuracy (validity) of this impression and how it has been formed:

> In the last 25 years, psychiatric historians have begun to redefine the framework of their investigations and to expand their use of source materials. Prior to the 1950s, work in psychiatric history was devoted largely to biographies of renowned pioneers in the field, surveys of early psychiatric theories, evaluated ahistorically in terms of their successful anticipation of 20th century ideas; and narrative, institutional histories of illustrious mental hospitals. More recent studies, by physicians and professional historians, have relied less on a "great man" theory of medical history to explain psychiatric progress and placed greater emphasis on the immediate historical and intellectual context in which psychiatric ideas evolved.
>
> ***Neugebauer, 1979:477***

Neugebauer's research took him to the archives of the Court of Wards (held at the Public Records Office in London[2], see Neugebauer, 1979:483 notes 26 and 27). There, he found records/documents of the very first mentally disordered individuals to have their needs provided for by legislation (this legislation being the *Prerogativa Regis*). The *Prerogativa Regis* distinguished between two types of disordered individuals: natural fools (or idiots) and those who were *non compos mentis* (later referred to as lunatics). The distinction was made on the basis of when the person became afflicted (from birth or later in life), and this had implications for how their estate could be used to care for them. As defined by the act:

> [An] Idiot is he that is a fool natural from his birth and knows not how to account or number 20 pence, nor cannot name his father or mother, nor of what age himself is, or such like easy and common matters; so that it appears he has no manner of understanding or reason, nor government of himself, what is for his profit or disprofit.
>
> ***quoted in Neugebauer, 1979:479***

> [A person is *non compos mentis*]...when any that before time hath had his wit and memory happen to fail of his wit...
>
> ***ibid.***

In other words, an idiot is someone disordered from birth, whereas a lunatic is only suffering a temporary loss of sanity. The provisions made by the act (under Sections 11 and 12), for the care of such individuals, were as follows:

> The king (sic) shall have the custody of the lands of natural fools taking the profits of them without waste or destruction and shall find them their necessaries....And after the death of such idiots he shall render it to the right heirs, so that such idiots shall not alien nor their heirs...be disinherited.
>
> ***ibid.***

> Also, the King shall provide, when any that before time hath had his wit and memory happen to fail of his wit, as there are many per lucida intervalla that their lands and tenements shall be safely kept without waste and destruction and that they and their household shall live and be maintained competently with the profits of the same, and the residue besides their sustenation shall be kept to their use, to be delivered unto them when they come to right mind; so that such lands and tenements shall in no wise be aliened and the King shall take nothing to his own use.
>
> ***ibid.***

This meant that the Crown could keep the *profits* made from the estate of an idiot, but *not* those made from the estate of a lunatic. It was, therefore, in the family's interest to have

2. The Court of Wards exercised its authority through the Chancery Court. It was set up in the late 13th century and continued until 1646, when the civil war forced it to cease operations. In 1660 its authority was officially transferred to the Chancery Court, where its contemporary equivalent (the Court of Protection) can be found to this day (Neugebauer, 1979:477-8).

the disordered individual declared a lunatic rather than an idiot. Here, the courts seem to have simply accept the word of the relative(s) (Neugebauer, 1979, 1989). If the person was suffering from lunacy, a relative (or responsible individual) could petition the courts to become the person's legal guardian (Neugebauer, 1979, 1989). This meant that they (rather than the Crown) could manage the estate of the disordered individual (until they recovered). In order to be recognised as the person's legal guardian the relative had to petition the courts. Many of these petitions include the relative's explanation for the person becoming disordered (although not a legal requirement as such). These petitions can now be found in the archives of the Court of Wards and the Chancery Court, at the Public Records Office in London (Neugebauer, 1979). Neugebauer has made a study of these explanations, a number of which can be found in his 1979 paper. The following table details of some of these explanations.

Case of:	Petition Date	Explanation
John Fitzwilliam	*July 1490*	*"gravely ill"*
John Norwich	*1502*	*"he has lost his reason owing to a long and incurable infirmity"*
John Gore	*2nd June 1626*	*"an illness which he had before that time"*
Bartholomew de Sakevill	*18th September 1291*	*"blow received on the head"*
Robert Barry	*1366*	*"induced by fear of his father"*
James Benok	*18th August 1568*	*"he became...afflicted by reason of a fright on 20 October 1556 and has so continued from that time to the present"*
Henry Williams	*6th August 1629*	*"visited with extreme sickness being 72 years...the extremity of which...long and tedious sickness, with want of natural rest has much weakened and debilitated his head, in so much that he is in great fear of lunacy"*
Richard Bowater	*11th May 1620*	*"a burning fever by means whereof and his old age* [70 years old]*...he sometimes is deprived of his memory and for the time becomes distracted"*
Unknown	*9th May 1631*	*"Distraction did rather proceed by some distemperature of diet and drink"*
Elizabeth Gate	*9th February 1622*	*"subject to a deep melancholy and discontent...by reason of her husband's death and his sale of all his lands to about £1,500 per annum and the care of her children and want of maintenance"*
Thomas Parsons	*29th November 1620*	*"Out of a desire to augment his estate for the better maintenance of himself, his wife and family...entering into consideration with himself how himself and his poor wife and family might hereafter be relieved and his own debts satisfied, and finding no means to accomplish the same for want of such debts as were due to himself, he became discontent and troubled in mind in so much as he grew weak in judgement....Since which time he is become mad and lunatic...he being now in Bedlam"*
William Burrishe	Unknown	*"put in great fear by thieves breaking into his mother's house in the night...became distracted and speechless most times"*
Edmund Franklyn	*17th November 1617*	*"melancholy sickness and some touch of remorse of conscience"*

After Neugebauer, 1979:481-482

Neugebauer's argument is not that everyone in the middle-ages had an appreciation for physical and psycho-social stressors, but that not everyone believed in super-natural causes. The evidence, presented in his paper(s), is not enough to overturn the current theory (mis/conception), but it does challenge it. This is why the operationalising of one's research question can be so important. Neugebauer knew that the evidence would be too scarce and too fragmented to offer an alternative to the existing theory – but it would be enough to raise doubts about its acceptance at the time. This brings us back to some of the ideas about theories and evidence, which I discussed in *'Interacting with Statistics'* (Taylor, 2007). Theories are simply idea supported by evidence: if the evidence for one theory survives and the evidence for another does not, then the first will prevail! The idea that it only takes one exception to disprove a theory, implies that it is a easier (takes less work) to disprove a theory than it is to create one. From the above table, it is possible to deduce that Neugebauer would have had to have search through at least 340 years (from 1291 to 1631) of court records, just to find these few fragments of evidence (and the paper from which this comes is only seven pages long). Neugebauer's work is a good illustration of how the ideas, discussed above, work in practice. It shows how theories are dependant on evidence; how academic movements can influence the research design; and how to design research for a particular research question (Taylor, 2007).

Personal Documents – Incidental Records

A number of researchers have raised questions about the authenticity, validity and representativness of personal correspondence (Hakim, 1993, Platt, 1981a, 1981b). This is, no doubt, one of the reasons why 'personal correspondence' has not been widely used by researchers (Plummer, 1983). The first thing to say about these observations is that they appear not to recognise the different types and uses of 'personal correspondence', not to mention the different ways in which this information may, or may not, be valid. The terms 'authenticity', 'validity', and 'representativness' can mean different things, depending on whether the correspondence is researcher elicited or just everyday correspondence. Here again, one can see the influence of Positivism; and, here again, one can see researchers conceding to Positivist ideals. Given the earlier discussions, on this subject (in the sections on ethnography and *researcher elicited* personal correspondence – see chapters two and three respectively), it perhaps goes without saying that the relevance of the these 'criteria' depend on the methodology being used, the research question, and the focus of the analysis. Again, these are part of the underlying research design, which is discussed in detail in Taylor, 2007. (Although it is important to show how these concepts can be applied to different research designs, there is little to be gained from discussing the same points, in depth, every time.)

One of the best uses of personal correspondence can be found in the work of the Chicago School (Thomas and Znaniecki). At the time (1920ish), there were large numbers of Polish émigrés living in America: Thomas and Znaniecki believed that they could learn something about these people from the letters they wrote to, and received from, their families in Poland. They put an advert in the Chicago Journal, offering to buy people's letters for between 10 and 20 cents each. The result was a publication (*'The Polish Peasant'*) consisting of 764 letters from over fifty families (Thomas and Znaniecki, 1918-20, reprinted in 1958). To this day, *'The Polish Peasant'* is still regarded as one on the defining works of the Chicago

School (see Plummer, 1983). This design goes a long way to satisfying the criteria discussed in the above paragraph (authenticity, validity, and representativeness) – there is no reason to doubt the integrity of the source (authenticity), there is no reason to doubt the sincerity of the information (validity), and the sample is large enough (764) to represent a reasonable cross-section of Polish émigré society (representativeness). It is, however, worth bearing in mind that Thomas and Znaniecki were investigating the lived experiences of a particular *social group*. (As sociologists they would have seen things on a macro-social level.) This meant that they *had* to have a large, representative sample, which helped to satisfy the other two criteria. It is also possible to analyse personal correspondence in terms of what they tell us about the author, the author's personality, and the author's perspective on life. In other words, a more psychological approach. (Note: there is nothing here about the validity of the contents, and you will see why later.) The examples, we are going to look at, are from individuals who are either eccentric or insane.

Before Edward VIII fell in love with, and married Wallis Simpson, he was in love with another (married) woman – Mrs. Freda Dudley Ward. The Duchess of Windsor's estate contained a collection of 263 (love) letters written by Edward VIII (then Prince of Wales) to Mrs. Ward. The collection has been published in a volume by Rupert Godfrey entitled *'Letters from a Prince'*. We will be looking at just two letters from that collection. Before doing so, let us look at what the editor (Godfrey) had to say about the couple. According to Godfrey, Mrs. Ward "clearly liked the attentions of the men who clustered around her…and she was not averse to encouraging the attentions of more than one admirer at a time." (xvii) His opinion of Edward VIII is even less flattering: "It is sometimes hard to believe that the writer of these letters was the heir to the greatest throne in the world: his immaturity and insecurity show themselves constantly, and, even without the benefit of Freda Dudley Ward's replies to him, it is clear what a tremendous support she was to him. She must have had extraordinary patience and tolerance to be able to put up with his unceasing self-pity and unhappiness with his lot." (viii) Godfrey's characterisations are a good starting point for our analysis. The first, and most obvious, question to ask is: *'to what extent do we agree with Godfrey's characterisations?'* From this one might ask: *'what do the letters tell us about Edward, Prince of Wales, and the sort of King he would have made (or even if he really wanted to be King)?'* The answers to these questions can clearly be seen in the material below. (I will leave you to make you own evaluation.)

Example 9:
Letter from the Duke of Windsor to Mrs. Freda Ward (1918)

British Mission with the Belgian Army,
Brussels
12th December 1918

My own very darling sweetheart,

….We went to look at some marvellous Hun submarine shelters at Bruges which are protected by feet of concrete from bombs; no-one but Huns would ever have thought of that & done it as it must have cost millions!! We also looked at a

Hun 15" naval gun near Ostend which has shelled Dunkirk etc. & the amount of construction for that is amazing!!...

On Tuesday night we dined at the Palace Hotel full of whizzbangs [fast young women: 'whizzbang' was trench slang for a mortar or small artillery round] with whom we all danced & then I went on to a nightclub with Claud & some Canadian friends where we stayed till 4:00...

You must forgive your E as it's nearly 3:00 & he's been dancing more at the Palace Hotel after dining with the King & Queen of the Belgians at their palace; my uncle went with me & it wasn't fearfully stiff though we were there nearly 3 hrs instead of 2 hrs as we had calculated & I got so so bored!!

Of course this dancing at the Palace Hotel isn't much fun though it's quite amusing...the 'whizzbangs' are rather revolting anyhow, very common, but a few of them can dance in a sort of way & there are 2 good string bands...of course all these women are 'bochesses' & you can guess what that means, darling, & really life in Brussels is practically the same as when the Huns were here, we 'officers allies' have merely replaced the Hun officers & we use Hun paper money!!...

These Belges aren't any happier with us than they were with the Huns; Gud! how I hate them though not as much as the Dagoes!!

There are lots of good restaurants & shops here though the prices of everything are ruinous if not prohibitive...you would be amazed at some of the 'additions' such as 1 whisky & soda 10 francs & eggs [gap] francs each!!

I'm off tomorrow to join Australian Corps at Ham...My brother returned to England with H.M. on Tuesday to get some more 'clothes' or some such feeble excuse; a curious boy in some ways so that I'm glad to have this chance of talking to him like a father, which I've been able to do several times this last fortnight!!

After Godfrey, 1998:117-8

Example 10:
Letter from the Duke of Windsor to Mrs. Freda Ward (1919)

Buckingham Palace,
SW
2nd May 1919

My very own darling beloved Fredie-Wedie,

...It's such ages since we've been separated darling that I've quite got out of practice in writing letters...

I just loathed seeing you drive away this evening, though how that last divine *'petit regard et sourire'* out of the taxi window cheered me up!!...

I hope your mama isn't terribly ill & that by the time you get this she will have pulled through her operation all right. But what sickening bad luck that it should all have had to happen this weekend when we had planned out everything so marvellously well. Still, it was the right thing for you to do, to go up to your

poor Mama, one must think of one's family sometimes as I have to, particularly one's Mama!!
I've had a terribly sordid evening & am not thrilled one atom at having been made a freemason; Christ, what I think of it all, particularly heavy men's dinners when I have to get up & make a speech...
I came in ½ hr ago & have been reading over my speech for tomorrow night's Academy dinner [the Royal Academy]...I haven't seen Bertie [Albert, later George VI] as he hasn't come in & don't expect him to either till Christ knows when in the morning!!
Tonight's stunt was all very weird & mysterious though there are really no secrets to give away...
What a divine fortnight this last one has been & I've never been so happy... I'm just longing to hear your sweet little voice on the telephone when you'll be able to tell your David [a name also used by Edward VIII] when he's going to see you again...
I must finish this letter tonight...& I'm so afraid of missing the post, which would make me cry & you 'thulky', & I just couldn't bear to think my little Fredie was 'thulking' when I wasn't there to 'unthulk' her!!!! What babies we both are sweetheart, but I do love it so & it does us both so much good, doesn't it?....
I just love your babies darling & think them divine & I'm so longing to see them again next week....
'Pleath' oh! 'pleath' come back to your devoted & adoring *petit amoureux* your little David soon & make him really happy all over again...

After Godfrey, 1998:143-4

The final example we are going to look at requires some explanation. It is a postcard written by someone suffering from schizophrenia: specifically, disorganised type[3] schizophrenia. This is far to complicated to explain in depth, but, for the purposes of this demonstration, it is sufficient to just know the diagnostic criteria (for this condition), which is as follows:

"(I)ncoherence, marked loosening of associations, or grossly disorganized behavior, and, in addition, flat or grossly inappropriate affect (emotions/moods). There are no systematized delusions..., although fragmentary delusions or hallucinations, in which the content is not organized into a coherent theme, are common."

The American Psychiatric Association's Diagnostic and Statistical Manual of Mental Disorders, Third Edition Revised, 1987:196

3. This letter is reproduced in Carson, Butcher and Coleman (1988: 333) as an example of paranoid schizophrenia. The diagnostic criteria for paranoid schizophrenia (according to the American Psychiatric Association's Diagnostic and Statistical Manual of Mental Disorders, Third Edition Revised (DSM 3R) are: a preoccupation with one or more systematized delusions...and None of the following: incoherence, marked loosening of association....(1987:197). Carson et al. say that one can see '"loosening" and deviations of thought' in this example (1988: 333) and one can. Therefore it can not be 'paranoid' schizophrenia according to the DSM 3R criteria. A more appropriate diagnosis would be 'disorganised' schizophrenia, for which the diagnostic criteria are: incoherence, marked loosening of associations, or grossly disorganised behaviour.... (DSM 3R, 1987:197).

To: The football department and its members present and future
The University of New Mexico, Albuquerque, N. M.

I depend on correct, honest supplementation of this card by telepathy as a thing which will make clear the meaning of this card. There exists a Playing of The Great Things, the correct, the constructive, world or universe politics, out-in-the-open telepathy, etc. According to the Great Things this playing is the most feasible thing of all; but it is held from newspaper advertising and correct, honest public world recognition, its next step, by telepathic forces (it seems), physical dangers, and lack of money. Over 10,000 cards and letters on this subject have been sent to prominent groups and persons all over the world. Correct, honest contact with the honest, out-in-the-open world. This line of thought, talk, etc. rule. The plain and frank. Strangers. The Great Things and opposites idea. References: In the telepathic world the correct playings. Please save this card for a history record since it is rare and important for history.

Acknowledgements to Carson, Butcher and Coleman and Scott, Foresman & Co (Allyn & Bacon).

The symptoms of the disorder can clearly be seen in this correspondence – the ideas fit together very 'loosely', and are also slightly delusional. What is clear, however, is that the person is desperately trying to communicate something, which they believe to be important. It is ironic, and perhaps a little tragic that this person has asked for their letter to be kept for future reference. Their letter *has* been kept, but not for the reasons they thought it would be.

Personal Documents – Intentional Records

This final section is about documents that are intended to be a record of a person's life (or some aspect of it). The documents are organised according to the amount of information they convey about the person's life. They range from the least informative, which are photographs; to the most informative, which are autobiographies. In this context, *'the amount of information they convey'* is purely impressionistic: it is simply a guide to the organisation of the section. Given the aims and objectives of the book, this discussion will focus on the reasons for the production of such documents, and what they show, rather than how they have been used, by researchers, in the past. As has already been mentioned, validity (in the form of impression management) can be a problem with this type of source. This often involves looking for reasons as to why the author may want to create a particular impression of themselves. To this end, Langlois and Seignobos have produced the following guidelines:

> All that interpretative criticism tells us is what the author meant; it remains to determine (1) what he really believed, for he may not have been sincere; (2) what he really knew, for he may have been mistaken. We may therefore distinguish *a critical examination of the author's good faith*, by which we seek to determine whether the author of the document lied or not, and a *critical examination* of his *accuracy*, by which we seek to determine whether he was or was not mistaken....

The questions in the first series will help us to inquire whether we have any reason to distrust the sincerity of the statement....

(1) The author seeks to gain a practical advantage for himself; he wishes to deceive the reader of the document, in order to persuade him to an action, or to dissuade him from it; he knowingly gives false information: we then say the author has an interest in deceiving....
(2) The author was placed in a situation which compelled him to violate truth. This happens whenever he has to draw up a document in conformity with a rule or custom, while the actual circumstances are in some point or other in conflict with the rule or custom....
(3) The author viewed with sympathy or antipathy a group of men (nation, party, denomination, province, city, family), or an assemblage of doctrines or institutions (religion, school of philosophy, political theory), and was led to distort facts in such a manner as to represent his friends in a favourable and his opponents in an unfavourable light....
(4) The author was induced by private or collective vanity to violate truth for the purpose of exalting himself or his group. He made such statements as he thought likely to give the reader the impression that he and his possessed qualities deserving of esteem...
(5) The author desired to please the public, or at least to avoid shocking it. He has expressed sentiments and ideas in harmony with the morality or the fashion of his public; he has distorted facts in order to adapt them to the passions and prejudices of his time, even those which he did not share....
(6) The author endeavoured to please the public by literary artifices. He distorted facts in order to embellish them according to is own aesthetic notions. We have therefore to look for the ideal of the author or of his time, in order to be on our guard against passages distorted to suit that ideal....

The second series of questions will be of use in determining whether there is any reason to distrust the accuracy of a statement....

(1) The author was in a situation to observe the fact, and supposed he really had observed it; he was, however, prevented from doing so by some interior force of which he was unconscious, an hallucination, an illusion, or a mere prejudice....
(2) The author was badly situated for observing. The practice of the sciences teaches us what are the conditions for correct observation. The observer ought to be placed where he can see correctly, and should have no practical interest, no desire to obtain a particular result, no preconceived idea about the result. He ought to record the observation immediately, in a precise system of notation; he ought to give a precise indication of his method. These conditions, which are insisted on in the sciences of observation, are never completely fulfilled by the authors of documents....
(3) The author states facts which he could have observed, but to which he did not take the trouble to attend. From idleness or negligence he reported details which he has merely inferred, or even imagined at random, and which turn out to be

> false. This is a common source of error, though it does not readily occur to one, and is to be suspected wherever the author was obliged to procure information in which he took little interest....
> (4) The fact stated is of such a nature that it could not have been learned by observation alone. It may be a hidden fact – a private secret, for example. It may be a fact relating to a collectivity, and applying to an extensive area or a long period of time; for example, the common act of a whole army, a custom common to a whole people or a whole age, a statistical total obtained by the addition of numerous items. It may be a comprehensive judgment on the character of a man, a group, a custom, an event.
>
> ***Langlois and Seignobos, 1898:165-76***
> ***emphasis in the original***

If researchers are only going to use information which they know, for certain, is valid, then they are going to end up rejecting most of the information in these kinds of documents. All too often, researchers will reject a source of information, on the grounds that it may have been (or has been) distorted, without actually considering what those *distortions* say about the author, their motivation for producing the document, or its function. With this in mind, let us now look at the different types of documents.

Photographs

Photographs were once thought to be 'incorruptible': the photographic process, itself, necessarily ensuring that the image could be nothing but a true representation of the subject. However, with the advent of digital photography, the photograph is no longer the 'honest' document that it once was. Before digital photography, it was quite easy to see when a photograph had been 'doctored' (Mitchell,1994), but with digital imaging technology it is now almost impossible to tell whether, or in what way, an image has been altered. Digital imaging technology appeared in the early 1990s, but the computing power required to process the images/information made the technology too expensive for the domestic market. In recent years, there has been a spectacular increase in the processing power (and specifications generally) of domestic computers. This, together with a fall in prices, has meant that digital imaging technology is now within most people's budget.

In 1994, an article appeared in the journal *'Scientific American'* entitled *'When Is Seeing Not Believing?'*. It was one of the first to point out the implications of this technology for (the integrity of) photographic images. Underneath the title, the author (William Mitchell) issued this terminal proclamation: "Digital technology for manipulating images has subverted the certainty of photographic evidence" (1994:44). This begs the question: *'what use is this (photographic) 'evidence' to researchers now?'* If the question is interpreted as relating to the validity of the information, then all digital images should be regarded with suspicion. Rejecting all digital images, on the grounds that they may have been 'doctored', could be a little premature. While the image/representation may not be quite as 'true to life' as it could be, the document still tells us something about that author, the subject, and the relationship between the author and the subject. Namely, that the subject is something which the author wishes to record for posterity. (Although photography is now an art-form, and some subjects are photographed for their aesthetic qualities (or are given these aesthetic qualities by the medium), it is being discussed here in the context of a record of daily life.) Again, the answer to this question lies in the *function* of the document.

When photography first began, it was too cumbersome (and presumably expensive) to use for anything other than portraits and the occasional special event. As photography became cheaper and more widely available, 'portraits' evolved into family chronicles, and 'special events' no longer needed to be *all that* special: Christmases, birthdays, holidays, even days out, all became worthy of a photograph. At this point, social scientists began to recognise that (as a record of contemporary culture), photographs could be used for research purposes (see Plummer, 1983; Scott, 1990). Arguably, photographic technology has evolved precisely because photographs do provide a means of recording things how they really are. The problem with this, is that it is not, necessarily, how the person 'sees' the subject, *in their own mind*. It is probably the case that most people do not like having their photograph ('picture') taken. This is because the photographic process does not 'air-brush' (out) spots, wrinkles, grey hairs, cellulite, rolls of fat, etc. in the way that our mind does. (How many times have you heard people say "I look awful in that photograph", and yet (to you) they do not seem to look too bad.) Imagine what would happen if photographs *could* be more like the images people have of themselves – a possibility with digital imaging technology. This means that, in the future, photographs may not be images of people, as they are, but as they think of themselves. Not only this, but it is now possible to edit (mistakes and 'boring bits' out of) home/amateur videos, giving even this medium the potential to be an idealised record of events. In many respects, digital photography can be seen as the 'third age' of photography. (The first two being the 'portrait/special event' era, and the 'popular culture' era.) It is too soon to tell whether the manipulation of such images will follow any particular pattern (or patterns). If it does, then this will be a potential source of information, and/or area of investigation, for social scientists.

Curriculum Vitaes (CVs)

A CV is basically a record of one's education and employment history. Its main purpose is to provide prospective employers with an account of a person's skills and work history, such that they can make an informed decision as to whether or not that person is suited to the position for which they have applied. As such, this has caused some researchers to class it as an 'administrative' document (Miller and Morgan, 1993). Whether one classes a CV as a 'personal account' (as it is described here) or an 'administrative document', CVs follow certain conventions regarding the content, layout, and narrative of the information they convey. In terms of their content, CVs typically include: the person's name, their contact details, academic history /educational qualifications, professional training, employment history, any other relevant experience (skills and abilities outside of formal education, unpaid work experience, professional memberships, awards and distinctions, publications etc.) and interests. (This is also the order in which this information is, typically, given.) It is becoming the convention to list one 's academic history in chronological order and one's employment history in *reverse* chronological order (i.e. from the most recent backwards) – the assumption being a person will progress in their career, and that their most recent (current) position is the best indicator of their (current) ability to date. In many respects, this reflects the dual role of a CV, as both an account of one's education/employment history, and as the first stage of a job application.

Prospective employers expect a CV to tell them whether or not the person has the qualifications and experience to do the job (or to take their application further). The onus is on the prospective employee to show (the prospective employer) that they have the qualifications and experience to do the job. A prospective employer should not (and probably would not) have to search through a CV, in order to find the information they need. A CV, therefore, has to include not only the prescribed information (which it is hoped will be 'right' for the job), but it has to convey it succinctly (two sides of A4 is generally considered to be a maximum length). If a CV is difficult to read because it contains too many details, or is confusing, then it is not a *'good'* CV. Similarly, if a CV has too *little* information in it, then this may also make it difficult (for a prospective employer) to see whether the person is qualified to do the job. In this way, convention and function dictate not only the content and format of a CV, but also the level of detail.

Within this framework, there is scope for a certain amount of impression management, and it is this 'self-presentational' aspect (of a CV) which has interested academics (Miller and Morgan, 1993). Using Erving Goffman's work on *'The Presentation of the Self in Everyday Life'* (Goffman, 1959), Miller and Morgan have looked at CVs from just such a ('self-presentational') perspective. It is difficult to determine just how much can be learned from using such an approach (Goffman's), but Miller and Morgan make a number of interesting observations in the course of their analysis. For Miller and Morgan, the CV presents the reader with an *impression* of one's career, and part of that *impression* is the progress one is making in that career. According to Miller and Morgan, 'the impression of progress' (my terminology) is conveyed by the information *no longer* included on the CV. The 'omitted' information is that which the author believes can now be *taken as read* (assumed/taken for granted). By omitting this information, the author is effectively saying "I must know how to do this (or have done this) at this point in my career". The omission is intended to give the impression that one's career has progressed to the point where it is no longer necessary to evidence such 'basic' skills, and this tells us something about how that individual sees/presents themselves and their career. It is important to appreciate that, in this context, 'basic' is a relative (to one's current position) term: the information, which a person omits because they see it as 'basic' (or would like people to think that they see it as 'basic'), conveys the *impression* that this is where they are (or would like to be seen as being; or see themselves as being) in their career.

Temporary or 'holiday' jobs are unlikely to be included on a CV, because they are unlikely to be relevant to the person's *career.* (As you know, these jobs are usually taken to relieve the financial burden of a fledgling family or a university degree: they are done 'just for the money'.) Similarly, the inclusion, or omission, of certain hobbies can be used to strengthen the impression created by this 'career' information. A CV is not, therefore, just a record of a person's career, it is the means by which they can create an impression of themselves, *through* that career. While the evolution of a CV tells us a lot about the subject (person) and how that person presents themselves (Miller and Morgan, 1993), the problem with this type of analysis is that people rarely keep back-dated copies of their CV. (By definition, the omitted information belongs to an 'out-of-date' CV.) With only the current version to go on, it is almost impossible to tell what information has been omitted (or promoted). Alternatively, one could compare the CVs of different people, at different stages in their career (following the same career), but this may not be practical.

These three elements (convention, audience and impression) each contribute to the format and content of the document (CV). The format is determined by convention and audience (prospective employers), whereas the content is determined by convention, audience *and* (the) impression (one wishes to create). With documents of this type, it is usually possible to make a critical analysis (of the contents) from 'knowing' just the author's intentions and the 'target' audience. However, in this case, the document (a CV) serves a combination of at least two functions – as an application for employment, and to create a particular impression of oneself. Any such analysis may, therefore, depend on being able to 'deduce' the function to which the author has given priority (the author's intention). In other words, whether the document is intended for a particular position, or simply as an exercise in self-promotion. On the face of it, this may look as though it requires a knowledge of the 'facts' (the content of the CV). There is, however, another way to approach this problem, and that is to look at the difference between a CV sent in response to a particular position, and (ideally the same) one forwarded by way of a general introduction. The person specification (of the advertised position) may serve as an indicator of the degree to which the CV has been 'modified' (in order to meet the requirements of the position). While this design may appear to be self-serving, it would, nonetheless, show how these two functions effect the document. This is only possible because, here, (in the case of a CV) the two functions (practical and impression management) are quite distinct. (Unlike many documents, of this type, where they are virtually synonymous.)

Diaries

Researchers tend to see diaries only as a source of information. Consequently, the texts on this subject (the use of diaries in research) emphasise the importance of being able to evaluate the validity of the account (Platt,1981a; Plummer, 1983). The approach advocated here (in this text) is to analyse this information, *in the context of the author's reasons for producing the document.* The problem with this approach is that we know very little about the kind of people who keep diaries, or their reasons for doing so:

> Diarists rarely seem to record the reasons why they keep a diary, and so this has to be inferred from the contents of their writings or from whatever else is known of their lives....The diarist...will tend to have some strong practical or literary interest for maintaining a personal record for any period of time; and self-justification or the impulse to record what are regarded as important personal or family events might be expected to dominate the motivation of many diarists and autobiographers. People reflect on whole periods of their life and then construct, or reconstruct, diary entries in the light of the self-image that they have built for themselves. As people periodically reconsider themselves, so their perception of what to record and how to record it will change. As the project of keeping a diary becomes incorporated in a person's self-perception, so the diary may become more of a self-justificatory autobiography rather than a daily record of events.
>
> ***Scott, 1990:177***

The first thing to say about this quote is that it is a very loose statement from an otherwise scholarly text. John Scott is, undoubtedly, one of the foremost authorities on documentary sources, but he offers no evidence for this particular scenario. The elements of Scott's scenario are plausible enough: diarists are self-absorbed individuals, who keep diaries in order to record their thoughts and feelings, which results in a certain amount of self-reflection, causing them to re-interpret past events, through self-serving justifications – but other scenarios may be equally plausible. The reason for discussing it is that this scenario has implications for the kind of analysis suggested here (and any other form of analysis, for that matter). If Scott is correct, then it is not simply the case that 'the diary' (as a class of document) is subject to impression management, but that the impression management is a consequence of the diary being defined by, and through, its own self-reflection. In other words, a diary may start off as a record of daily events, but (through self reflection) it later develops into the diarist's 'legal' counsel. (Someone to whom the diarist can explain their actions, and make these actions appear justified/right/good/, in the eyes of their imaginary court.) This suggests that the 'self-presentational component' may not be a 'stable' part of the document (which makes it difficult to identify and, therefore, isolate). It is not clear whether diaries do become more self-serving, as they go on, but diarists may well oscillate between a state of honest reflection and self presentation. It may also be the case that, in the course of writing the diary, the diarist comes to realise something about themselves, which they change at a later date (Lee, 1993:117). The 'amending' of subsequent (or even existing) diary entries may be not so much an attempt to hide the 'truth' as an attempt to make the diary more consistent with the 'reformed' self.

This scenario is based on one rather important assumption – that 'the diary' is what it purports to be: a record events, made on a daily basis. Until recently, the diaries of Samuel Pepys were assumed to be just that (Plummer, 1983; Scott, 1990). Although there is little doubt as to the authenticity of Pepys' diaries (they were written by Pepys, at the time), there is, however, some doubt as to whether they were written in the way in which they purport to have been written:

> (W)hilst it is common to believe that Pepys simply "sat down every night for nine years just scribbling with effortless frankness the little incidents which [he was] honest enough to record as having caught [his] attention at the moment"...it becomes clear...that it was not produced daily, but rather evolved: "a product fashioned with some care, both in its matter and style."
>
> ***Plummer, 1983:18***

This 'revelation' calls into question one of the advantages of using diaries (as a source of information), over (say) autobiographies – the integrity of the information. The assumption being that, since the information is recorded while it is still fresh in the person's mind, it will subject to fewer distortions (by their memory) (see Coxon, referred to in Lee, 1993). (We will look at the implications of this, in more detail, in the next section.) Regardless of whether they were written daily, or retrospectively, (or even with considerable artistic licence) Pepys' diaries were, at least, written by the person they purport to have been written by – Samuel Pepys. However, authenticity is only an issue if the document is significant enough to make it an issue. In the early 1980s, a previously unknown researcher discovered the diaries of Adolph Hitler (and promptly offered them to a British newspaper, for **one million pounds**). The *'Hitler Diaries'* (as they became known) were declared 'authentic' by a "prominent British historian" (Scott, 1990:176). However, tests later revealed that the ink and paper were both modern (Scott, 1990). They were fakes!

Very little is known about the *kind* of people who keep diaries, or what distinguishes them from the people who do not. Scott suggests that "the methodical and introspective routines of diary-keeping may appeal mainly to those with a methodical and introspective character." (Scott, 1990:177) Diary-keeping may also appeal to those (such as Anne Frank or Samuel Pepys) who appreciate the importance of recording their experiences for posterity (Plummer, 1983). Scott's scenario seems to suggest that keeping a diary fulfils some kind of need in the diarist. Understanding what that need is, and why it exists, may (ultimately) be the answer to the question *'what kind of people keep diaries?'* This suggests that, like CVs, diaries have a practical function (the recording of certain daily events), and a self-serving function (the discussion of one's thoughts and feelings). This can clearly be seen in the distinction drawn (above) between those diaries kept by Anne Frank and Samuel Pepys, and those kept be individuals of a more "introspective character". As with the practical and self-serving functions of a CV, they may not be as distinct as this example implies. What is interesting about this is that CVs and diaries (as types of documents) both seem to have a practical and a self-serving function, but they take different *forms* depending on the nature of the document. This is also the case with biographies and autobiographies.

Biographies and Autobiographies

Biographies and autobiographies can also be seen to have a practical, and a self-serving function. By definition, they are accounts of a person's life, but they are told not just from a particular perspective (usually that of the person themselves), but with a particular theme. Common themes seem to be (from reading and watching many biographies): rags to riches, surviving adversity, a misjudged individual, a tragic life, a glamorous 'front', and so on. While it is sometimes obvious, and perfectly understandable, when someone sees (and want to portray) themselves as having one of these 'life types', it is not always obvious when a biographer is doing the same (to their subject) – McCann's (1988) biography of Monroe is a good example. Monroe's screen persona, together with circumstances surrounding her death, have contributed to her image (which sill lives on) as an attractive, but vulnerable, young woman. As such, she can still evoke the male instinct to protect and defend a vulnerable young woman. McCann's biography is more a declaration of his love/admiration, for Marilyn Monroe, than it is an account of her life. This is not the only way in which the image making system serves the biographers. Biographers, and autobiographers, often trade on the knowledge that we, the public, see only the star's 'screen image', or the image created, for them, by the film studios, (and we know it), promising us a glimpse of 'the person behind the image'. It would, perhaps, be naïve, of us, to think that these auto/biographies do just that. Aside from this issue of 'life typing', (what is left of) the reference material would have still have had to have been interpreted by the author. Even if the (auto)biographer's memory for events, (themselves) their subject, their (own) subject's life, and their subject's (own) family and friends, is infallible, and copiously documented, this 'evidence' is still subject to the author's 'reading' of it.

While biographies and autobiographies may be 'unreliable', as sources of information *per se* (particularly if the person is famous), the literature, which is currently available on celebrity image making, the film industry, and celebrity auto/biographies, makes it possible to deconstruct screen images in such a way that the biographical material can be understood from the perspective of those who constructed the image. However, this does not mean to say that the approach teaches us *nothing* about the person. As we will see, John Wayne was instrumental in cultivating his screen image, but he also thwarted the film studio's attempts to distort certain facts about his youth (in order to make them consistent with his screen image) (Levy, 1988). This tells us something about what kind of *person* he really was. In many respects, it is much easier to deconstruct the image/auto/biography of a celebrity, than it is to deconstruct the image/auto/biography of an 'ordinary' individual (because the construction process is very often documented/ evidenced). There is no shortage of material describing the way in which *particular* celebrity 'images' have been created, but there is very little documentation on the way in which this ('star') system works, *in general.* Commentators, on the subject, are in no doubt that it is definitely a *'system'*:

> In building a public personality the motion picture industry has perfected the device of stereotyping its stars. The star system is based on the premise that a star is accepted by the public in terms of a certain set of personality traits which permeate all of his or her film roles. The successful stars have been those whose appeal can be catalogued into a series of such traits, associations and mannerisms.
>
> ***Harris, 1991:40***

It is not difficult to find cases where this 'formula' has been applied to certain 'movie stars'. In Grace Kelly's case, there was "an emphasis on her family's adherence to the good life, particularly the drama of her father's achievements in sport and business." (Harris, 1991:42) Together with "the use of quotes from her co-stars about her off-screen lady-like qualities." (ibid.) These devices served to reinforce the stereotype created by her screen roles (ibid.). However, these techniques are not always directed towards the star's social background. On screen, Mae West was an intimidating and seductive 'woman-of-the-world', but, in reality, she was short and ungainly. When photographed in high heels and feathered gowns, she appeared to have a much fuller figure, and to be a lot taller than her five feet three inches (160cm). Her tendency to slouch was made less noticeable by having her walk more slowly on set. This slow, deliberate gate, also made her appear more seductive (Walker, 1966). It is often the case that biographers can agree on these 'image stereotypes', better than they can the facts of the person's life. (As in the case of Rudolph Valentino.)

> While...biographers rarely agree on any facts concerning Valentino's life, they stereotypically relate his personal success and suffering to the ongoing crisis of American culture and social roles.
>
> ***Hansen, 1991:259***

Although it is not always possible to know just how much of a 'celebrity image' is simply that *(an image)*, these examples show how 'celebrity images' can be deconstructed almost as easily as they can be constructed. Our interest here, however, is not so much the construction and deconstruction of celebrity images, as the deconstruction of celebrity images, in order to learn something about the person themselves. (Clearly, this approach lends itself to a variety of research questions.) By looking at the part played by celebrities, in the construction of their own image, and their attitude to this practice, one can often see the person behind the 'image'. This is clearly illustrated in E. Levy's biography of John Wayne (1988).

Illustrated Example John Wayne (Levy, 1988)
Although John Wayne, and his films, are known the world over, his screen career followed a very definite path, which is not immediately discernible. John Wayne did not so much play characters, as *role models*. These roles embodied, and epitomised, contemporary ideals of masculine behaviour – the display of moral fibre, together with a belief in family values. "Wayne's movies were obsessed with defining the parameters of masculine behavior, repeatedly using two devices: comparing him with women an the one hand, and with weak men on the other." (Levy, 1988:146) Even so, they still relied on Wayne giving a convincing performance. It is tempting to think (as most people, no doubt, do) that Wayne was simply playing himself in these roles. There is no denying that his range of roles was very narrow, but this did not stop him from becoming one of the highest paid actors of his day. This meant that John Wayne's public image was a hard act to follow – even for John Wayne.

When John Wayne's son first spoke, publicly, about his father, the cinema going public was surprised to learn just how easy-going John Wayne really was. (His daughters were not, however, of the same opinion.) This is not all that surprising given his record in high school, and his *entrée* into the film industry. As a young man, Wayne neglected his football scholarship in order to spend time at the local film studio. For some reason, he was attracted to the film industry, which (at the time) was not the most masculine of occupations! His (real) name (Marrion Michael Morrison) was not suited to the roles he played in his early career. Consequently, he was billed as *'Michael Burn'* or *'Duke Morrison'*. His nick-name, *'the duke'*, was not part of an attempt to make him sound more masculine; it was the name of his dog. (The people at the studio could remember his dog's name better than they could his – and so it became his nick-name.) Wayne's relationship with women was not that of his screen image either. His second marriage broke down amid allegations of physical abuse. (He was married, and divorced, a total of three times.) As an actor, he was *not* an overnight success. He began his career as a prop-man, and was only offered (small) acting roles because he was willing to take more risks than any of the other actors. In the 1930s he appeared in 67 westerns, most of which were, very forgettable B-Movies. His pay was just $7.50 per week. It was 1939 before he got his 'big break' (in the film *'Stagecoach'*).

We can only *speculate* on the *relationship* between the decisions he made in his film career and the events of his personal life. This can clearly be seen when attempting to 'deduce' the reasons for his decisions. For example, it is difficult to tell whether his 'trademark' screen image was a consequence of his failure to make it as a singing cowboy (see Levy), or a reaction against the (less than masculine) reputation of the acting profession, at the time. We can see, from the above, that he was not as macho as his screen image. While the allegations of abuse may well have been just part of the recriminations, which are now all too familiar in divorce cases, he did certainly seem to have difficulty understanding women. All of his (three) marriages failed for the same reason – he did not spend enough time with his family. (He was working on his films.) It may have been that he did not know how to express his feelings, and his career helped him to avoid these situations, or he may have simply believed that he *was* being a good father, by *providing* for his family. Either way, it looks like they would rather have seen more of him than of his films.

His son remembers him as being liberal, because he allowed him to pursue his own choice of career. Wayne neglected his football scholarship in order to pursue his acting career. Given the reputation of the acting profession, at the time, it is unlikely that Marrion Morrison's father would have approved of *his* (rather liberal) career choice. (Or perhaps he did!) John Wayne could not very well criticise his son for doing what he had done himself! His daughters, however, had a rather different view of their father. This is because he would not let them wear make-up until they reached a certain age; he would not let them smoke in the house until they were married; and he would not let them take drugs. Is this just an over-protective father, or a tyrant? It is probably the case that most parents would not approve of their children taking drugs. With regard to smoking, we do not know whether he was aware of (or suspected) the (health) risks involved (he suffered from cancer himself), or whether he just thought that he was responsible for them while they were still *his daughters*, and so made this decision for their own well-being. In would appear that, in his eyes, getting married made them adults, and capable of taking responsibility for their own decisions.

He was clearly determined to succeed as an actor; his reputation for hard work, and dedication, is widely acknowledged, and from what we know about his beliefs and values, he may well have preferred this to being an overnight success. While this tells us something about Marrion Morrison/John Wayne's life, and what sort of person he *may* have been, it tells us very little about what sort of person he was. It is only when we look at the part played by Marrion Morrison, in the construction (or debunking) of 'John Wayne', do we actually see something of the person. Arguably, Wayne saw his screen image, not as a type of person, but as the personification of certain moral values. Undoubtedly, he had standards, and tried to incorporate these into his screen/public image. He is, famously, quoted as saying: "my children will remember that I told them never to lie – to other people or to themselves" (in Levy, 1988:265). Rather ironically, these standards were often a *barrier* to the creation of that (desired public) image. A case in point is the way in which he reacted to the studio's attempts to 're-define' (lie about) his earlier life:

> Despite his enormous popularity, Wayne remained modest throughout his career. From the beginning, he refused to co-operate with the publicity machine of the studios, especially when he thought the image sold to the public was too contrived. For the publicity tour of 'The Big Trail', Fox fabricated the idea

> that he was a former Texas ranger and an All-American football star. Wayne debunked this imagery and told reporters that it was phoney. To make Wayne's nickname, Duke, more appealing, the studio said it stemmed from a part he had played in a high-school play. This was consistently denied by the star, "I was not named after royalty. I was named for my dog." For the publicity tour of 'The Big Trail' Fox dressed him in buckskins and gave him some props, but he quit in the middle of the tour because he felt ridiculous waving a hatchet on stage and aiming a muzzle-loading rifle. Fox also faked the idea that he knew how to ride a horse from his childhood – which many magazines 'bought' at the time – but Wayne later said he had never been on a saddle before he went to Hollywood.
>
> ***Levy, 1988:235***

This shows us the role that 'John Wayne' played (or refused to play) in the construction of his own public image. It is incidents, such as this, which give us a glimpse of the 'real' person. We can be confident about the validity of the information, and what it tells us about the person, *because* it shows the person acting contrary to, and to the detriment of, their screen/public image. At this point, it is perhaps worth reiterating the comments made earlier – about the construction process not always being this well documented. In some cases, the information may have to be 'pieced together' from a variety of sources. Nonetheless, this example *does* show that it is possible to obtain valid information from (even) the most fabricated of life stories.

Section Conclusion

Seeing personal documents only in terms of the information they contain, and the validity (or otherwise) of that information, is missing the point of what personal documents are actually for. Whether the document is created in the course of running a business, keeping in touch with friends, witnessing an important occasion, applying for a job, attempting to deal with one's conscience, or describing what kind of life one has had, its purpose is to record what the author wants to record. It is probably the case that the originator's main concern is *not* the validity of the information. Understanding what that main concern *is* (the function of the document, for the author), enables one to understand the nature of the information contained in such documents. Researchers seem to be quite critical of the quality of information contained in such documents (compared to the quality of the information in the designs they are able to control themselves), but one has to remember that the information is not intended to be of 'professional quality'; it is not intended to be for research purposes (for a discussion on the ethics of this approach, see Harrison and Lyon, 1993). The originators, of personal documents, are not professional record-keepers, and personal documents are not professional records: the information is for the originator's own personal use. At first sight, it may be difficult to see what oral histories have in common with such documents. (Oral histories being, essentially, accounts of the past, brought forth at the request of the researcher.) However, as we will see, oral histories can also be seen to serve a function for their 'teller'.

Personal - Verbal

Again, there is essentially only one source/method which fits this description – the oral history. Although it is possible to argue that most oral histories are elicited by researchers, given that they constitute a account of events, oral histories more closely resemble diary entries, or biography extracts, than they do interviews. It, therefore, seems more appropriate (given the typology) to discuss the method/source here (under the heading of *'Personal Accounts'*), rather than under the heading of *'Researcher Elicited Personal Information'*. While the concept of an oral history is relatively straightforward (the respondent recounts an event from their past), research suggests that the 'memory' for an event, and the act of 'recounting' it, may not be what they seem. This clearly has implications for the validity of the method. In order to understand what an oral history is, it is necessary to look at 'oral histories' in the context of this literature. Hence, this section will not only look at examples of oral histories, but the research into *episodic* memory and 'story' telling.

Oral Histories

An oral history is the recounting of an 'historical' event (usually at the request of the researcher), which the person witnessed during their lifetime. In this case, 'historical' does not necessarily mean 'historically significant' – it may simply be an anecdote about a way of life which has since passed. Irrespective of the historical significance of the event, the expectation is that the respondent will be a valid and reliable source of information, particularly if the event happened to them personally. (Reliable simply means that the person will be consistent.) Much of the literature, on this subject, (the use of oral histories in research) is, therefore, concerned with the validity of such accounts/information. There are, however, some situations where it may be the *only* source of information available on a particular subject, and this may mean accepting it as the *best* (source of information) *available*.

This tendency to see, and classify, oral histories as 'just' another type of 'documentary' evidence (i.e. 'like' the documentary evidence discussed above), ignores some very important differences between 'documentary evidence' recorded as a document, and 'documentary evidence' retained in someone's memory. The first, relates to the nature of memory itself. Human memory is not like the storage device of a computer – it does not record *everything*; it does not record it *precisely*, and it makes *inferences* where necessary. (Personal documents, on the other hand, record information much more meticulously.) The second, relates to what is 'memorable'. Most people can remember their first day at school, but few can remember what they had for dinner last week. (Personal documents often record information which would otherwise be forgotten.) The third, relates to the task of 'recounting'. 'Recounting an event' means telling someone what happened. This is not the same as recalling the 'facts'. It is often the case that respondents will want to tell an interesting story – *their* story. (This is clearly not the same as retrieving, and reading, a document.) Understanding these 'differences' enables one to understand the content and form (some may say the weaknesses) of an 'oral history' (as a source of information). As with other 'documentary' sources, oral histories should, perhaps, be seen in terms of what they tell us about (what is important to) the author, rather than as a source of information *per se*. Let us now look at this in more detail, starting with the nature of memory.

The Nature of Memory

Since 1932, researchers have had evidence that 'the memory' is not simply a 'place' where information is 'stored' and 'retrieved'. This evidence came from a series of experiments conducted by a gentleman who's name is now synonymous with our current understanding of 'memory' – Sir Frederick Bartlett. Bartlett was interested in the way in which 'memories' become distorted over time. His research simply involved giving participants a story, and then asking them to re-tell it at a later date – the errors being evidence that our 'memory' is not simply a storage/retrieval device. Bartlett realised that he could not use a story with which participants were, or may have been, familiar (because the participants may have simply been recalling what they had heard previously). He, therefore, decided to use a tale from Native American folklore – *The War of the Ghosts.* Bartlett's resign reveals something quite interesting about the nature of 'memory' – the way in which events are interpreted and reconstructed. This can clearly be seen in the following example, which shows Bartlett's original story, together with the versions re-told by one of his participants (at different points afterwards).

Example 11:
Sir Frederick Bartlett's 'The War of the Ghosts' (1932)

The War of the Ghosts Story

> One night two young men from Egulac went down to the river to hunt seals, and while they were there it became foggy and calm. Then they heard war-cries, and they thought: "Maybe this is a war-party." They escaped to the shore, and hid behind a log. Now canoes came up, and they heard the noise of paddles, and saw one canoe coming up to them. There were five men in the canoe, and they said: "What do you think? We wish to take you along. We are going up the river to make war on the people".
> One of the young men said "I have no arrows".
> "Arrows are in the canoe", they said.
> "I will not go along. I might be killed. My relatives do not know where I have gone. But you," he said, turning to the other, "may go with them."
> So one of the young men went, but the other returned home.
> And the warriors went on up the river to a town on the other side of Kalama. The people came down to the water, and they began to fight, and many were killed. But presently the young man heard one of the warriors say: "Quick, let us go home: that Indian has been hit." Now he thought: "Oh, they are ghosts". He did not feel sick, but they said he had been shot.
> So the canoes went back to Egulac, and the young man went ashore to his house, and made a fire. And he told everybody and said: "Behold I accompanied the ghosts, and we went to fight. Many of our fellows were killed, and many of those who attacked us were killed. They said I was hit, and I did not feel sick".
> He told it all, and then he became quiet. When the sun rose he fell down. Something black came out of his mouth. His face became contorted. The people jumped up and cried. He was dead.
>
> ***Bartlett, 1932:65***

This is how one person (Subject P) re-told the story over a period of three years:

First re-telling (shortly after hearing the story):

> Two youths were standing by a river about to start seal-catching, when a boat appeared with five men in it. they were all armed for war.
> The youths were at first frightened, but they were asked by the men to come and help them fight some enemies on the other bank. One youth said he could not come as his relations would be anxious about him; the other said he would go, and entered the boat.
> In the evening he returned to his hut, and told his friends that he had been in a battle. A great many had been slain, and he had been wounded by an arrow; he had not felt any pain, he said. They told him that he must have been fighting in a battle of ghosts. Then he remembered that it had been queer and he was very excited.
> In the morning, however, he became ill, and his friends gathered round; he fell down and his face became very pale. Then he writhed and shrieked and his friends were filled with terror. At last he became calm. Something hard and black came out of his mouth, and he lay contorted and dead.

Second re-telling (two weeks after hearing the story):

> There were two young men who once went out in the afternoon to catch seals. They were about to begin when a boat appeared on the river and in it were five warriors. These looked so fierce that the men thought they were going to attack them. But they were reassured when they asked the youths to enter the boat and help them to fight some enemies.
> The elder said he would not come because his relations might be anxious about him. But the other said he would go and went off.
> He returned in the evening tired and excited, and he told his friends that he had been fighting in a great battle. "Many of us and many of the foe were slain", he said. "I was wounded but did not feel sick."
> Later in the evening he retired quietly to bed, after lighting a fire. The next morning, however, when the neighbours came to see him, he said that he must have been fighting in a battle of ghosts.
> Then he fell down and writhed in agony. Something black jumped out of his mouth. All the neighbours held up their hands and shrieked with terror, and when they examined the youth they found that he was dead.

(The third (six weeks after hearing the story) and fourth versions are very similar, which is why the third has been omitted.)

Fourth re-telling (three and a half months after hearing the story):

> Two youths went down to the river to hunt for seals. They were hiding behind a rock when a boat with some warriors in it came up to them. The warriors,

> however, said they were friends, and invited them to help them to fight an enemy over the river. The elder one said he could not go because his relations would be so anxious if he did not return home. So the younger one went with the warriors in the boat.
> In the evening he returned and told his friends that he had been fighting in a great battle, and that many were slain on both sides.
> After lighting a fire he retired to sleep. In the morning, when the sun rose, he fell ill, and his neighbours came to see him. He had told them that he had been wounded in the battle but had felt no pain then. But soon he became worse. He writhed and shrieked and fell to the ground dead. Something black came out of his mouth. The neighbours said he must have been at war with the ghosts.

Fifth re-telling (almost three years after hearing the story):

> Some warriors went to wage war against the ghosts. They fought all day and one of their number was wounded.
> They returned home in the evening, bearing their sick comrade. As the day drew to a close, he became rapidly worse and the villagers came round him. At sunset he sighed: something black came out of his mouth. He was dead.
>
> ***After Bartlett, 1932:73-75***

The way in which our memory is thought to work (as evidenced in this demonstration) is as follows. Although we experience life in all its detail, only the 'gist' of those experiences are committed to memory (Bourne and Ekstrand, 1985:182). What we are (actually) recalling, when we recall of those experiences, is the 'gist' of the information, reassembled into a coherent narrative. We can see this in the above example – the participant clearly remembered the 'gist' of the story, but their memory 'narrated' it slightly differently each time. It is interesting to see how different aspects of the story were remembered at different times. It also interesting to see (in the above example) how ideas were borrowed, from elsewhere, in order to make the story makes sense. The participant (/participant's memory) clearly felt the need to explain why the young men were venturing on to the river. This is how the idea of a seal hunting party came to be introduced into the story. It is interesting to see how different aspects of the narration, not only became part of the 'memory' (for the story), but were, themselves, 'remembered' and re-narrated differently. The episode, where the main character 'shrieked', was never part of the original story, and yet (through progressive re-telling of the story) this was first 'remembered' as an act of the main character, and later as an act of the neighbours. This raises some very important questions about the nature of 'memory': *'Is it possible to be mistaken about some of the original details of an event?' 'Could someone believe that a 'reconstructed' memory was, in fact, a 'true' memory?' Or even: 'Could someone be made to 'remember' events which never happened?'* The answer to all of these questions is 'yes', and the (empirical) evidence is quite persuasive.

In 1973, Johnson, Bransford and Solomon conducted an experiment in which they gave subjects the following sentences:

> John was trying to fix the birdhouse. He was pounding the nail when his father came out to watch him and help him do the work.
>
> It was late at night when the phone rang and a voice gave a frantic cry. The spy threw the secret documents into the fireplace just 30 seconds before it would have been too late.

They then asked participants to confirm whether, or not, the following sentences were what the heard:

> John was using the hammer to fix the birdhouse when his father came out to watch and help him do the work.
>
> The spy burned the secret documents just 30 seconds before it would have been too late.

A significant number of participants (incorrectly) claimed that the second sentence was the one that they actually heard (cited in Bourne and Ekstrand, 1985:182). The first two sentences did not mention the words 'hammer' and 'burned', but these can be considered logical inferences (given the nature of the episodes). The importantly finding, here, is that some participants actually believed (incorrectly) that they had heard something, which was different to what they had actually heard. If we look at the narratives, then the second set of sentences definitely present the more coherent and comprehensible story-lines. (More like those which one's memory would come up with?) A similar effect was observed, by Laurence and Perry (1983), in what was a very different kind of investigation. Laurence and Perry hypnotised participants into believing that they had been woken by a noise in the night. Almost half of their participants (13 out of 27) accepted the suggestion, some of whom reconstructed the evening's events in order to explain why they 'awoke' during the night. This is just one example:

> "I'm pretty sure it happened because I can remember being startled. It's the physical thing I remember…I'm making an assumption that it was a noise but I was conscious of the different cars. It must have been something like that. I can remember the startle."
>
> ***Laurence and Perry, 1983:524***

Even when they were told that the memory had been implanted, by hypnotic suggestion, some of them simply refused to believed it. In the words of one participant: "I'm pretty certain I heard them. As a matter of fact, I'm pretty damn certain. I'm positive I heard these noises." (Laurence and Perry, 1983:524) This shows that not only can memories be induced, but that confidence in a particular memory is no guarantee of its validity. It can, of course, be argued that this is hardly surprising given the 'power' of hypnosis.

The last example, in this (sub-)section, comes from a paper by Elizabeth Loftus. Loftus is best known for her work into eyewitness testimony (as evidence in court cases), but more recently, she has begun investigating a phenomenon known as *False Memory Syndrome*. In a paper, published in 1993, she discussed the case of teenager, who is helped to remember an incident from his past (nine years earlier). This is how his older brother remembered the incident:

> "It was 1981 or 1982. I remember that Chris was 5. We had gone shopping at the University City shopping mall in Spokane. After some panic, we found Chris being led down the mall by a tall, oldish man (I think he was wearing a flannel shirt). Chris was crying and holding the man's hand. The man explained that he had found Chris walking around crying his eyes out just a few moments before and was trying to help him find his parents."

Over the next few days 'Chris' gradually began to recall the event:

> "That day I was so scared that I would never see my family again. I knew I was in trouble....I remember mom telling me never to do that again....I also remember that old man's flannel shirt....I sort of remember the stores....I remember the man asking me if I was lost."
>
> ***Loftus, 1993:532***

In this case, it is difficult to know which account is the more accurate. The older brother's memory would have, presumably, been better, but the younger brother did remember something which the older brother did not: being reprimanded by his mother. This suggests that the two boys may have been remembering what was personally relevant to them. The problem is that (and this is a problem with all oral histories and/or eyewitness testimonies) we do not know what actually happened (hence these 'witnesses' become our source of information). However, in this case, we do know what happened, and the answer is: none of it! It was part of an experiment, by Loftus, to see whether false memories could be implanted without using hypnosis. (The older brother was acting under the direction of the researcher.) This research comes in the wake of several cases in which 'victims' of child abuse have been able to retrieve 'repressed memories' with the help of 'suggestions' from a psychotherapists. In many cases, this testimony was enough to incriminated their 'abuser' (usually a male relative). Research now suggests that some of these 'incidents' may never have happened (see Loftus, 1993 for a discussion).

On the face of it, it would appear that human memory is too fallible (as a recording device) to be of use to researchers. However, let us look again at what this literature actually shows. In the study on hypnotically induced memories, less than half of the participants accepted the (hypnotic) suggestion – and this was only for a single, non-specific detail. The 'false memory' literature (Loftus) shows that memories can be induced, but it would appear that the person has to be *trying* to come up with something (usually over a period of weeks). In the case of psychotherapy, a person is motivated to 'remember' the incident, in order to deal with their emotional problems. Similarly, in the Loftus example, the older brother's ability to remember the incident placed an implicit expectation, on the younger brother, to remember the incident as well. In other words, induced memories seem to require 'effort', if not a protracted attempt to 'find' (create) the memory. Arguably, these findings have far greater implications for the (research) design/procedure, than they do for the information itself. They suggest that researchers should not put too much pressure on participants (to 'remember'), because this is likely to force them to come up with *something*: perhaps creating a *false* memory of the event? The 'mistakes' made in the Johnson, Bransford and Solomon study were only with respect to the *'details'* of the incident (and these were consistent with the incident, anyway). Similarly, the 'mistakes' made in the Bartlett study, were more 'interpretative' than 'substantive'. This means that memories recalled, in oral histories, are unlikely to be 'false' *(per se)*, and that 'mistakes' are likely to be consistent with the narrative of the incident. While this tells us what to expect from our memory, it does not tell us what kind of events are likely to be remembered.

Memorable Events

The nature of memory is such that our memory for an event not only includes the 'facts', it also includes how we feel at the time. Bower, Monteiro and Gilligan (1978) conducted an experiment in which participants learned, and recalled, lists of words, in different emotional states (induced by hypnosis). The researchers concluded that:

> An intense emotion aroused by an event may firmly lodge that event in memory because the emotion may be distinctive and seldom experienced at this intensity. Moreover, personal emotional events tend to be thought over repeatedly, with the result that the event becomes associated with the persisting mood state (e.g., grief). The associations may be dormant for many years until the intense emotional mood is re-experienced, at which time the person may "spontaneously" come forth with memories of earlier incidents associated with the same mood.
>
> ***Bower, Monteiro and Gilligan, 1978:584-5***

This has been put forward as an explanation for the self-sustaining nature of depression - "Once someone is depressed they will find it easier to recall depressing memories than happy ones, and this will deepen the depression." (Smyth, Morris, Levy and Ellis, 1987:228)

Emotions, therefore, are not simply 'recorded' with the event; they can cue certain memories, and may even be part of the mechanism for remembering, itself. This forms the basis of, what is referred to as, the *'Now Print'* hypothesis (Livingston, 1967). Although rather dated, the *'Now Print'* hypothesis proposes that highly emotive situations trigger our brain to record what is present at the time.

> The 'Now print!' hypothesis proposes that, when an event is recognized as being highly novel or unexpected and is seen as being of biological meaning, then lower brain areas involved in emotions (the limbic system) massively activate cortical areas causing all recent brain events to be recorded in memory. The term 'biological meaning' is meant to convey the idea that some events directly impinge upon the durability of the self. In evolutionary terms, we could imagine a special memory mechanism developing which served to record, in some detail, features of events which were markedly life threatening or, maybe, life enhancing. These details would then be available to the encoding organism for reflection.
>
> ***Conway, 1990:163***

The *'Now Print'* hypothesis (as an explanation for 'flash bulb' memories) is only weakly supported by the evidence (see Conway, 1990 for a discussion), but the idea on which it is based is *plausible* enough. While the 'memories', for such an event, may not be as detailed as the theory would have us believe (ibid.), it certainly seems plausible to suggest that a highly emotive event/episode may also trigger a mechanism for remembering it. The argument being offered here (in this text) is that highly emotional events are 'memorable'. In other words, 'memorable' events are memorable *(for that person)* because they are highly emotional. It is also stands to reason that highly emotional events will be *personally* significant. In many respects, this brings us back to the self-serving function of documentary evidence. As we will see in the next (sub-)section, recounting the event means telling *their* 'story'.

Recounting the Event

In Bartlett's *'War of the Ghosts'*, the first thing that the warrior did, when he returned home, was to tell his family about the unusual incident. This is what people do when something unusual happens to them (or they witness something unusual) – they tell someone about it! In this respect, the *'War of the Ghosts'* story, and Bartlett's paper, tell us something about the nature of memory; but the story *itself* also tells us something about how people react when something out of the ordinary happens (and this is obviously not restricted to our own culture). The more unusual (or emotionally significant) the event, the more likely people are to talk about it. This also ensures that the memory of it will be kept 'alive' (Conway, 1990). It is hardly surprising, therefore, that almost everyone (who was alive at the time) can remember where they were, and what they were doing, when (John F.) Kennedy was shot. The same appears to be the case for those who were alive when Lincoln was shot. This is an account from one such individual (recounted 33 years after the incident):

> My father and I were on the road to A- in the state of Maine to purchase the 'fixings' needed for my graduation. When we were driving down a steep hill into the city we felt something was wrong. Everybody looked so sad, and there was such terrible excitement that my father stopped his horse, and leaning from the carriage called: "What is it my friend? What has happened?" "haven't you heard?" was the reply – "Lincoln has been assassinated." The lines fell from my father's limp hands, and with tears streaming from his eyes he sat as one bereft of motion. We were far from home, and much must be done, so he rallied after a time, and we finished our work as well as our heavy hearts would allow.
>
> ***Reported in Colgrove, 1899; cited in Conway, 1990:62***

Before discussing this extract, let us look at what Martin Conway (an expert, working in the field of autobiographical memory) has to say about it:

> It has the feel of a story which has been told many times and which has been embellished with literary devices such as "the lines fell from my father's limp hands", "he sat as one bereft of motion", etc. The structure of the account

> carefully builds up to the shocking news and then tails off into the resulting sadness.
>
> ***Conway, 1990:62.***

Clearly, the respondent does not see it (recounting the incident) in terms of simply reporting the 'facts'; they see it in terms of an(other) opportunity to tell *their* 'story'. (It perhaps goes without saying that this is not a story in the sense of a fabrication; but a story in the sense of an *'interesting'* anecdote.) Which is why it is told with a *story-like* narrative. As far as the respondent is concerned, it (recounting the event) is an opportunity to be the centre of attention ("Guess what happened to me the other day?…"), but as far as the researcher is concerned, it is an opportunity to learn about the past. ("Tell me what it was like when….") This means that the respondent may be approaching the exercise in a way which is different to that which the researcher would like them to do (without either of them actually realise it). Researchers should, perhaps, accept that it is the *respondent's* 'story', and they are the one telling it.

A Reassessment of Oral Histories:
The nature of human memory, the distortions which occur when information is recalled, and the embellishments that are added in the narration, all raise questions about the validity of the information imparted in oral histories. While this means that oral histories may not the window on the past which historians would like them to be, it does not mean that they are of absolutely no value to social researchers. Arguably, oral histories should be seen, and utilised, for what they are – accounts of events, which are remembered *because* they are important *to that person.* In other words, looking upon an oral history not as an historical document *per se*, but as a way of understanding that person, their life, and (perhaps) the times in which they lived, by their recounting of an event that is 'memorable' to them. In this respect, it does not matter how 'valid'(precise) the account is because, *at the very least*, it tells us something about the life and social history of *that person* (and perhaps others living at that time). This means that, *in spite of the problems with validity*, researchers can still learn a lot from oral histories. The examples (of oral histories), reproduced below, are all quite different, and yet they all tell us something about that person, why the event was memorable for them, and how they felt at the time. Here again, we can see that the issue of validity is, perhaps, less of a 'problem' than some academics would have us believe. (These points are self-evident with the following examples.)

Example 12:
Oral History – from: 'Voices from the Third Reich' by Steinhoff, Pechel and Showalter (1991)

A Bavarian woman (Hilda Rubin), recalling an incident which happened during the second world war.

> As you can imagine, we were all excited to see my brother, Gunther, suddenly standing at the front door one day in 1945. But it didn't take long for us to realize he had come home illegally. He had had travel orders to Berlin and had decided

to make a side trip home to see us. He was 16 and a half years old and had been drafted a year and a half before. We hadn't seen him during that entire time he was homesick.

My mother was absolutely terrified. We were all scared that he was going to be arrested. Only a short time before, a young man from our neighbourhood had returned home and hidden in his parent's house. We found out about it later when he was arrested and shot as a deserter. Of course, we feared that the same might happen to my brother. It was impossible for him to leave from the train station in our town. He would have most certainly run into the hands of the many MPs (Military Police) there who were looking for deserters.

We decided my brother should leave from a small station about five miles away. That night, my sister and I got out our bikes – my brother sat on the beck of mine – and rode over the bumpy paths. We were really relieved when we reached the little village and dropped him off. The tiny waiting room in the train station was a sad sight. The place was filled with boys, all around 16 or 17, who had also made a quick visit home and wanted to avoid the MPs.

My mother was almost happy that my brother was gone again. 'Now at least nothing can happen to him. He can't be arrested and won't be shot,' she said. After that, we didn't hear anything from him. It was as though he had evaporated into thin air. We received no mail. My mother had a terrible time dealing with the fact that she had forced her only son to return to the war. Time and time again she said, 'We should have hidden him, we should have just kept him here!' At that time, no one knew what was the right thing to do anymore."

Steinhoff, Pechel and Showalter, 1991:335-336

Example 13:
Oral History – from: 'Thurcroft a Village and the Miners' Strike' by 'The People of Thurcroft' (1986)

A woman's first experience of picketing (standing at the entrance of a place of work, in support of a dispute or industrial action).

"The first time we went picketing we were really chuffed and proud. Afterwards we came walking back into the Welfare Hall but nobody wanted to know 'They must be silly buggers, fancy going out picketing.' That was my impression of what they thought. It was the same with the lasses who hadn't wanted to go. They said 'Oh have you?' and just carried on what they were doing, they weren't interested to know what had gone off. The union wasn't amused at all, I'll always remember seeing – 's face. I think actually what it was, they were concerned that if we got picked up, who would look after us. Also it was a man's world, let's face it, only men ever did things like that."

'The People of Thurcroft', 1986:174

Example 14:
Oral History – from: 'Ageing and Life History' by Coleman (1991)

An elderly woman ('Mrs Manners') reminiscing about her past.

> "...the days when we could get out and walk a lot in the countryside before the advent of cars, that was a lovely time. I do go now on the rare occasions that my son-in-law takes me to the beauty spots, but the beauty of it for me is spoilt by so many cars and charabancs (coaches) and that kind of thing, and I remember it of course when there wasn't, but this must be with progress, I mean I do realise that. More people can now enjoy it. It's just that when you've seen it, you know, as it used to be, it's not quite so good now, but as I say that's rather a selfish view, because so many more people now can enjoy these places."
>
> ***Coleman, 1991:139***

5: Research Methods and Sources – Case Studies & Published/Public Information

Introduction

The methods and sources, discussed so far, have only been considered in the context of a *'one method, one construct'* design. While this is, perhaps, the best way to demonstrate the various methods available to researchers, it does not accurately represent this particular stage of the research process (designing research to answer the research question). In practice, this stage of the research process is more complex than just choosing *the* method, or source, which enables one to best answer the research question. The designs considered, in *this* chapter, represent an advance on the above approach, in that they either use a *number* of methods, or they require a much more sophisticated understanding of the source. The first part of the chapter will be devoted to case studies. This represents an advance on the above type of design, because of its focus – a social unit rather than a social construct. The second half of the chapter will be devoted to published/public sources. This represents an advance on the above type of design, in that it requires a more sophisticated critique of the data. Publication often implies that the volume has been written by 'an authority' on the subject, and is, therefore, an authoritative source of information. We will see why this may apply to some sources more than others. The purpose of discussing such designs, here, is (the same as with the other discussions) to demonstrate how they can be used to answer certain types of research questions. (Although they may not be appropriate for a first project.) This will be sufficient to demonstrate the principles involved, but, given the nature of this text, it will be no more than a brief discussion.

Holistic	**Case Studies**
Published/Public Written	**Archives and Publications**
Published/Public Verbal	**The Media**

Holistic

Case Studies

Introduction

At this point, we have looked at methods of eliciting 'personal' information about individuals, and various sources of personal information. Having seen how these designs can be used to investigate facets of the personal, or social world, as general **constructs**, the next step is to look at how these can combined to investigate a particular single personal, or social 'unit' **(a case)**. As it is being used here, the term 'personal, or social unit' means any personal, or social, construct, with distinct temporal, social, or geographical parameters (borders). This can be anything from a period in someone's life to a large-scale organisation. The defining characteristic, of a case *study*, is that the research question is directed towards an investigation of a unit rather than a construct. In (the first half of) the next chapter, we will look at (two examples demonstrating) how researchers have used (some of) these methods (among others) to investigate a particular personal, or social 'unit'.

We have seen (in the preceding chapters) how it is possibly to apply a range of methods (and sources) to the task of *'answering the research question'*. Although it may not have always been clear, these (research) questions have related to **constructs** (concepts). Therefore, the *principal* design decision has been to choose which design (method(s) and/source(s)) will *best* enable one to answer the research question (as appropriate for this construct). With a **case study**, the focus of the investigation is not a construct, but a **personal**, or **social 'unit'.** This means that the 'research question' is, in part, an explanation of why it is important to study this *particular* personal or social 'unit'. Therefore, the *principle* design decision (to conduct a 'case study') can also be seen to depend on why it is important to investigate this particular personal, or social 'unit'. Once a researcher has made this *(principle)* 'decision', they then have to decide which methods are most appropriate for this particular case study (a *secondary* design decision). Here, 'the answer to the research question', and 'the decision to study a particular personal, or social 'unit'' are one and the same. This means that those methods, 'which best enable the researcher to answer the research question', are those methods, 'which best enable the researcher to have a *complete* understanding of the subject' (this particular personal, or social 'unit'). The decision is exactly the same, except (in this case) the methods are chosen to give an understanding of a particular personal, or social 'unit', rather than a construct. In other words, personal, or social 'units' often become 'case studies' because there is a reason to study them as such. The research question, design and subject are, therefore, 'three sides' of the same coin!

All of the above *('methods and sources')* discussions have two components: a methodological/philosophical component (the rationale behind the design) and a practical component (the application of the design to a particular type of research question): this discussion (of *'case studies'*) will be no different. The nature of the (case study) design does, however, mean that they have a slightly different form, in this discussion. Looking at it (case study research) from a practical point of view, case studies involve: a) identifying an *appropriate* 'case'; b) identifying *why* it is important to study that 'case'; and c) identifying the most *effective way* of studying that 'case'. Looking at it (case study research) from a methodological point of view, researchers (who intend to use this 'method') need to know: a) how to define/identify 'a case'; b) the reasons for choosing a 'case study' design; and c) the methods, which one can use in a 'case study'. With regard to the methodological component, the discussion will begin by attempting to define 'a case', in the context of a 'case *study*'. (The literature, on this subject, is not as clear as it could be, which makes some kind of unifying definition essential.) The discussion will then go on to look at the three principle reasons for choosing a case study design. Finally, the application of the (case study) design will be demonstrated using two, quite different, examples (at the end of the section). With regard to the practical component, the definition (of 'a case', and 'a case study') used here, can serve both as guide to one's own research design, and as a way of recognising other 'case studies'. The reason for choosing a particular personal or, social 'unit' can only be justified by reference to that 'unit'. While it is possible to identify the reasons for choosing a case study *design*, it is not possible to identify the reasons for choosing a case study *subject*. (This is effectively the research question, which could be anything.) Again, the same 'rules' apply with regard to methods and sources – some methods will be more appropriate (for investigating this subject) than others. Most of the methods referred to, in the demonstration examples, will either be familiar (from the above discussions), or require a minimal explanation.

Definition and Design

Many of the texts, on this subject (case study research), (for example, Yin, 1989; Hakim, 1992:Ch 6; and Mitchell, 1983), describe how to conduct a case study, and case studies which have been conducted by other researchers, but they do not describe the characteristics of a case study *(design)* in a way which allows researchers to know, for certain, that their design *is* a case study. Or even if an investigation can *rightfully* be called as a 'case study'. The 'authorities', on this subject, seem to be uncertain as to what constitutes a 'case study', or how to distinguish it from other types of research design. Many of the investigations, which are discussed (in the above texts) as examples of 'case studies', are also discussed (elsewhere) as examples of other methods. One reason for this is that case studies can, and do, make use of a range of research methods/sources, which means that they *are* examples of these (other) methods. However, this does not help researchers, when it comes to identifying the characteristics of the design *per se*. Similarly, some (research methods) texts have chapters devoted the combining of different research methods (Hakim, 1992:Ch 10; Burgess, 1993:Ch7), and this is often what researchers do in a case study. However, a case study is more than just a design, which uses a number of methods. Another reason for 'non-case study' designs being 'confused' with case studies has to do with the status afforded to case studies. A 'case study' implies that it is a comprehensive, in-depth investigation (certainly more comprehensive and in-depth than (say) a participant observation). Claiming that one's research is a 'case study' will, therefore, enhance its (and the researcher's) academic status. For this reason, researchers may claim that an investigation is a 'case study', when (technically) it is not. Hence, the distinction between case studies and other designs can, often, become 'blurred'.

Another 'barrier' to identifying the characteristics of the ('case study') *design*, as a single unitary design, is the way in which these texts focus on the *'different types'* of case studies. Clearly, focusing on the differences, rather than the similarities (between case studies), is not going to help, when it comes to identifying the characteristics of *a* case study *design*. These 'classifications' do, however, provide the basis for choosing a case study design. (Although, not all the categories can be used in this way.) To this end, these texts classify case studies in three ways – by depth, by design, and by unit of investigation. Mitchell (1983) uses the system of classification based on Gluckman's *'apt illustrations'*, *'social situations'* and *'extended case studies'* (repeated observations made over a period of time). Yin (1989) uses a system based on the function of the investigation: single critical case studies, single revelatory case studies, single embedded case studies, and multiple case studies. Hakim (1992) organises her chapter into – individual case histories, community studies, case studies of social groups, case studies of organisations, and case studies of events, roles, and relationships. (In this case, it is perhaps a little unfair to describe the structure of Hakim's chapter as a 'typology', because the chapter needed to have a structure, and this quite a good way of doing it.) Within these typologies, one can identify the reasons for choosing a case study *design*: 'apt illustrations', 'single critical case studies', and 'single revelatory case studies' roughly translate to: 'a good illustration of…', 'a turning point in…', 'a rare, unique, or exceptional case, from which we can learn something'. With regard to the choice of level/subject, this can be found in Hakim's chapter. These typologies, therefore, identify the *foci* and the reasons for conducting a case study, but they do not identify the characteristics of a case study design. Arguably, the 'mistake' which these authors are making is to try to 'deduce' the rationale of the design from the end-product of the (various) investigation(s). Case studies are considerably more 'layered' (more design decisions) than any of the designs considered so far. This makes it very difficult to discern the rationale of an investigation, let alone the rationale of the design *per se*. Not only this, but it is also questionable whether anything can be gained from differentiating between 'types of' case studies, anyway. Methodologically speaking, a 'case study' should have a set of characteristics, which define it as such, and set it apart from other (research) designs – distinguishing between different 'types' of case studies, only makes it more difficult to see what a *case study* actually is. (It is also questionable whether all of the examples, cited as 'case studies', are *actually* case studies.)

An alternative (approach) is to re-construct the rationale from first principles – the bottom up, which is the approach which will be taken here. A case study implies that *something* has been 'singled out' for investigation. The subject of a case study is, therefore, something which can be 'marked out' as a 'unit' of study. It must, therefore, be possible to identify this 'unit' in such a way as to distinguish it from its (social) (back)ground and/or set it apart from 'others'. The ability to mark out this 'unit', as a discrete entity, also implies that it can be studied *in isolation*. This means that it must be 'bounded' and distinct from its background. It should also be possible to identify why it is important to study this particular personal or social 'unit'. Or alternatively *'what can be learned from studying the subject, in this way (by a case study)?'* (You will recall that 'being scientific' means making the 'knowledge creation' process 'visible', so that it can be critically appraised by the academic community, see Taylor 2007 – which is what we are doing here.) Clearly (as has already been pointed out), these questions are related, if not inter-dependent. These questions also point to how the subject can best be investigated. (In other words the methods to use.) To this end, the term 'case study' also implies that it is a *complete study* of the *'unit'* (subject). In order to achieve this, the investigation must be both comprehensive and coherent (comprehensive, to be 'a complete study', and coherent, to be 'a study of the unit'). This means that the various aspects of the investigation must not only fit together as an explanation of the whole, but they must facilitate the understanding of each other, as elements of that whole. There should be no 'loose ends', no unanswered questions and no need for further explanations. It is possible to distil this down into, what may be considered, a definition of a 'case study':

> *The study of a particular personal, or social unit, in order to arrive at an understanding of that unit, as a 'whole'.*

Before looking at some examples of case studies (as defined above), it is important to think about the quality of the information. Yin devotes a certain amount of his book (on case study research) to discussing issues such as reliability and validity. (As you will, no doubt, recall – reliability means that the research remains current over time, and validity means that the research shows what it purports to show.) Given the nature of a case study, it is difficult to see how any (well designed) case study can not be valid, in the sense of it not being a study of what it purports to be. If the study is an investigation of a particular social unit, then this is what it is! The investigation can only be done well, or done poorly – it can not be done without validity (in this sense). A case study can, however, be 'invalid' in the sense that it does not provide a complete and coherent understanding of the subject (as a whole). In other words, it is invalid *as a case study*. A case study is, clearly, not going to serve its purpose, if it does not include enough information. It is also the case that too much detail may distract one from seeing the subject 'as a whole'. This association, between the design and validity, means that validity is effectively a measure of the quality of the investigation. In terms of internal validity (the design being consistent with the research question), the subject of the 'case study' must, by definition, be consistent with the reason for studying it. Given that this aspect of the design must, so obviously, be right or wrong, it is difficult to see how anyone could consider it an issue. It is, however, possible that, if the 'case' has been chosen as (say) an 'apt illustration', then it may not be an 'apt illustration' – but this is a problem which besets any form of sampling.

Similarly, it is difficult to see how external validity (the ability to generalise the findings to practical applications) can be that much of an issue, either. Given that the purpose of a case study is to investigate a particular 'unit', (if the investigation is done well) everything that one needs to know about the subject (that particular unit) will (should) be in that investigation. A case study is what researchers refer to as an **idiographic** approach. In other words, it attempts to produce an in-depth understanding of a particular subject. This is in contrast to a **nomothetic** approach, such as a survey, which aims to produce a broad, but relatively superficial, understanding of the subject. If the intention is to generalise the findings to other (similar) 'units', then a case study may not be the most appropriate method to use. For the same reason, reliability is not really an issue. A case study is designed to give an in-depth view of a particular subject, *at that moment in time*. Most case studies are about social aggregates (groups, organisations etc.), and social aggregates do not remain static. If another researcher were to re-investigate this 'case', then the differences (if any) are more likely to reflect a real change in the subject, rather than an absence of reliability in the original investigation. If the research question calls for a repeated investigation of the subject, then one wonders why it has not been designed as a quasi-experimental investigation. Although reliability and validity are issues in research, they are not *really* issues in case study design; they are either intrinsic to the design, or irrelevant by virtue of the methodology. It would appear that yet another researcher (Yin) has found it necessary to justify the 'scientific' credentials of their design, according to the standards laid down by Positivism. As with many of the (more idiographic) designs, discussed above, this defensive stance (with regard to reliability and validity) has caused researchers not to look at the *strengths* of the design, or to consider what the method was *actually* designed *for*. With this in mind, let us now look at some examples of *actual* case studies. The first is the case of an individual with a rather unique 'disability', and the second is the case of a public house with a rather dubious reputation.

Recovery from Blindness – The case of S. B. by Gregory and Wallace (1963)

This is a case study of one of the first people to be 'successfully' treated for cataracts. Today, we take this treatment for granted, but at the time (in the 1960s), it must have seemed like nothing short of a miracle. The advent of this treatment posed an intriguing question: *'what it was like to see the world through the eyes of a man (S. B.) who had been blind for most of his life (but could now see)?'* This case has been discussed, by the authors, many times (Gregory and Wallace, 1963; Gregory, 1974). The following discussion is based on the (six page) summary, which can be found in Gregory's *Eye and Brain* (1986:194-199). With an investigation of this type, the researcher has to be careful not to turn the person into an 'exhibit'. (It is very easy trap for researchers to fall into.) This means framing the 'research question' in a way, which maintains the respect and dignity of the individual. For this reason, Gregory presents S.B. as a person who can teach us something, rather than as a 'specimen' to be studied. This is how he introduces S.B:

> This case, a man of fifty-two, whom we may call S.B., was when blind an active and intelligent man. He would go for cycle rides, a friend holding his shoulder to guide him; he would often dispense with the usual white stick, sometimes walking into things such as parked cars, and hurting himself. He liked making things, with simple tools in a shed in his garden. All his life he tried to picture the world of sight: he would wash his brother-in-law's car, imagining its shape as vividly as he could. He longed for the day when he might see, though his eyes had been given up as hopeless, so that no surgeon would risk wasting a donated cornea. Finally the attempt was made (when cornea 'banks' were set up) and it was successful. But though the operation was a success, the story ends in tragedy.
>
> ***Gregory, 1986:194***

The first thing to notice about this 'biography' is that it is not just a collection of facts; it is a *'story'*. (You may notice a resemblance, here, between this and the 'stories' told in oral histories and auto/biographies.) This is more relevant than it may first appear. In the previous section, it was argued that case studies should constitute a complete and coherent account of the 'subject' (as a whole), and some researchers may interpret this to mean 'telling the subject's story'. Here, Gregory has fused the methodological with the literary to create this opening scene. As literary scholars would have no doubt noticed, this paragraph has the narrative of a tragedy. A tragedy is defined as *'a story, whose central character is intent on a course of action, which will ultimately bring about their downfall'*. To paraphrase Gregory's introduction – here is a man, who has longed to see for all his life, and just when it seems like he will never be able to see, fait steps in a the very last moment. The operation is a success, but joy turns to tragedy, when the operation (which he has wanted for so long) does not bring him the happiness he deserves. We pity him for wanting what brought him so much unhappiness – and this is what makes it a tragedy! Not only does Gregory know how to write a good tragedy; he also knows how to keep us in suspense. The observant reader may have notice that (in this introduction) Gregory does not *actually* say what happens to S.B., merely that the 'story' ends in tragedy. In order to find out what happens to S.B., we have to read on – and who would not want to, with an introduction like this? This sense of tragedy can also be seen in the, rather ironic, title given to the original study – '*Recovery* From Early Blindness' (1963, emphasis added).

Having 'introduced' S.B. (the 'case' and 'reason') the next task was to determine what he was 'seeing' (the 'method(s)' and 'data'). For this, Gregory chose to conduct a series of (what researchers would call) naïve experiments and observations. These are not the highly 'prescribed' experimental and observational investigations, discussed in chapter two, but a few simple 'activities' designed to (subtly) find out what the world was like for S.B. (They are experiments (and observations) in the sense that the situation has been 'set up' by the researcher for the purposes of gathering information, but they are not as rigorously controlled as a formal experimental (or observational) investigation, which is why they are referred to as naïve.) For example, one thing they (Gregory and Wallace) did was to take S.B. on a tour of London. It was 'research', but it was also something that they thought he (S.B.) would enjoy. They also used some of the objects, from that excursion, as the stimuli for a series of drawing exercises. In one such exercise, they asked him to draw an elephant, and then asked for his reaction when he saw one – he was not surprised! They also observed his progress (a series of attempts over time), in drawing a double-decker bus. (A bus/coach with two levels of seating.) There does, however, appear to be a period when he had to associate the visual images with a tactile sensation, in order to be able to 'see' the object. At one point, he said that he needed to touch a lathe in order to "see" it (Gregory, 1986:198). This also explains why he had difficulty judging distance, and learning letters by sight. The 'study' also includes a number of 'anecdotal' observations. For example, when he spotted an object that he had not seen before, and was surprised to learn (from one of the hospital matrons) that it was a quarter moon: he thought that a quarter moon would look like "a quarter piece of cake" (Gregory, 1986:196). We are also told that he became fascinated with reflections, and would sit for hours (in pubs/bars) watching people's reflections.

However, not all of S.B.'s experiences were positive or amusing. The prospect of crossing the road (on his own) terrified him. Prior to the operation, he was unable to see the danger, for himself (and was happy to put his faith in the person guiding him), but, after the operation, he could see how dangerous it was. This is why it often took two people to coax him across the road (Gregory, 1986:197). His new world disappointed him in many ways:

> (H)e (S.B.) became seriously dispirited...he found the world drab, and was upset by flaking paint and blemishes on things. He liked bright colours, but became depressed when the light faded. His depression became marked, and general. He gradually gave up active living, and three years later died.
>
> ***Gregory, 1986: 194-5***

Apparently, depression is a common reaction in such cases (and understandable, if you put yourself in their position) – as Gregory explains:

> Depression in people recovering sight after many years of blindness seems to be a common feature of the cases. Its cause is probably complex, but in part it seems to be a realisation of what they have missed – not only visual experience, but opportunities to do things denied them during the years of blindness. Some of these people revert very soon to living without light, making no attempt to see.

> S.B. would often not trouble to turn on the light in the evening, but would sit in darkness.
>
> ***Gregory, 1986:195***

When it comes to evaluating a case study, we evaluate them in the same way as we would any piece of research – *by how well the investigation answers the research question.* In this case, the research question is: *'what is it like to see, after being blind for most of one's life?'* Here, we can see how, with a case study, the research question, subject, and design justification are inter-related – the subject suggests the research question, which suggests the design. Furthermore, the episode/individual is bounded by time/physicality, and can be differentiated from its/their (back)ground, thus making it definable as a 'case'. The investigation also tells us how the researchers went about answering the research question (the methods used). Although replication is not *really* an issue here (with a case study), it is still important to be able to evaluate the way in which the data has been collected. As far as answering the research question is concerned, the investigation does tell us what it *feels like* to see for the first time (after being blind for most of one's life). The study (necessarily) can not tell us what that person 'sees', only what surprises and fascinates, frightens and disappoints them, about a world which, up to now, they have only 'seen' in their mind. In this respect, it has answered the (research) question.

Crime and Community – The case of 'The Grafton Arms' by Janet Foster (1990)
This is the (case) study of a sub-group of patrons, frequenting a particular South London public house (pub). 'The Grafton Arms' (as it is referred to by the researcher) is known to be the haunt of some rather roguish characters, and it is this (sub-)group of individuals which the researcher chose to make into a case study. Unlike the previous example, the parameters of this case require a little more explanation. In this respect, the first thing to say about the 'subject' is that both the establishment, and its clientele, can be marked out from their physical/social 'background':

> Surrounded by two large and notorious council estates, in a generally depressed area of South London, one would have been unlikely to wander into the Grafton Arms by chance. Despite a large ethnic minority population the pub's clientele were almost exclusively white and working class.
>
> ***Foster, 1990:20***

However, this only tells us about the location of the pub and its patrons. The subject of the investigation is a particular sub-group of patrons, which makes it necessary to identify and differentiate between different groups of patrons (frequenting this public house). The patrons of the pub were found to fall into a number of (social) groups, and these groups could be found in different areas of the bar:

Clientele of the Grafton

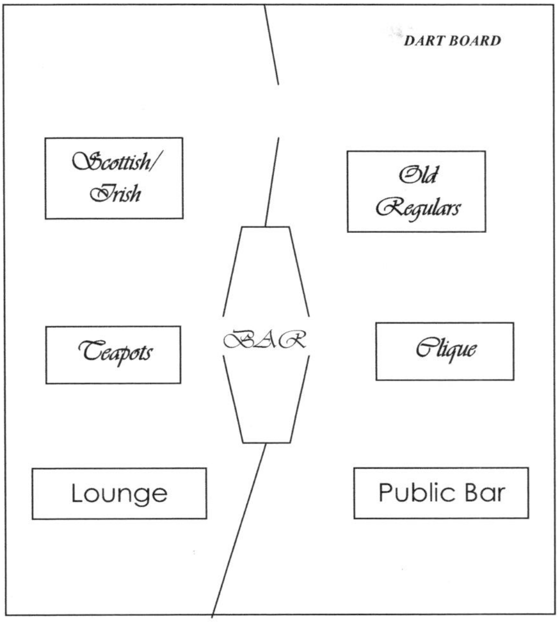

After Foster, 1990

Acknowledgements to Janet Foster and Routledge

Most of these labels are self-explanatory. The only label/group which is not self-explanatory is 'The Teapots'. The following description (of them) was supplied by one of Foster's informants:

> They were …born and bred here. Most of the Teapots are brothers…there's four or five brothers and they've just evolved around that….There's only ten of 'em I s'ppose and they're very close knit…I don't really know their full background but I think they've had a spot of 'bovver' at one time or another, which makes 'em that close knit…they very much look after each other. If one's in trouble they're all in trouble or the rest will get the other one out of trouble….If you're a Teapot you're a Teapot one hundred per cent.
>
> ***'Del' quoted in Foster, 1990:23***

The sub-group, which became the subject of this investigation, constituted its own social network (as well as most of the ladies and men's darts teams). Here is a list, and brief description, of those individuals (by darts teams):

Ladies' darts team:

Michelle (32)	Captain (Scottish)
Sue (28)	Michelle's sister
Chris (37)	Responsible for recruiting most members
Sal (37)	'Fetched in' by Chris (friends since their teens)
Vera (40)	Rarely played but drank with Chris, Sal and Carol
Carol (33)	

George (22)	(Because of her boyish appearance) Del's sister
Ruth (22)	Worked with George
Sandra (19)	Recruited by George
Men's darts team:	
Harry (mid-30s)	Captain
Del (37)	
Barry (35)	
Jimmy (34)	
Billy (28)	
Joe (31)	
Frankie (34)	
Paul (19)	The only youngster to play in the team, Paul was a talented player who spent most of his time with Jimmy, but also played pairs in another pub (which was run by one of Arthur and Vera's daughters) with Chris.
Scottish Dave (40s)	The only Non-Londoner in the team

After Foster, 1990:25

Notice, here, that Foster was able to gather more information about the ladies' darts team than the men's darts team. All too often, a researcher's gender will limit them to certain social interactions, which, in turn, will limit the information available to them. Quite naturally, Foster seems to have spent more time with the women. Given the social background of the group, and the fact that many of the men were married (or related) to the women (in the group), Foster was probably wise not to show too much interest in the men. (Just in case her intentions became misconstrued.) These individuals were part of a very tight-knit social network, and this is something, which Foster was also able to 'map'. (This mapping of social relationships is what is known as a **Sociometric** analysis.)

Social Networks and Relationships in the Grafton Arms

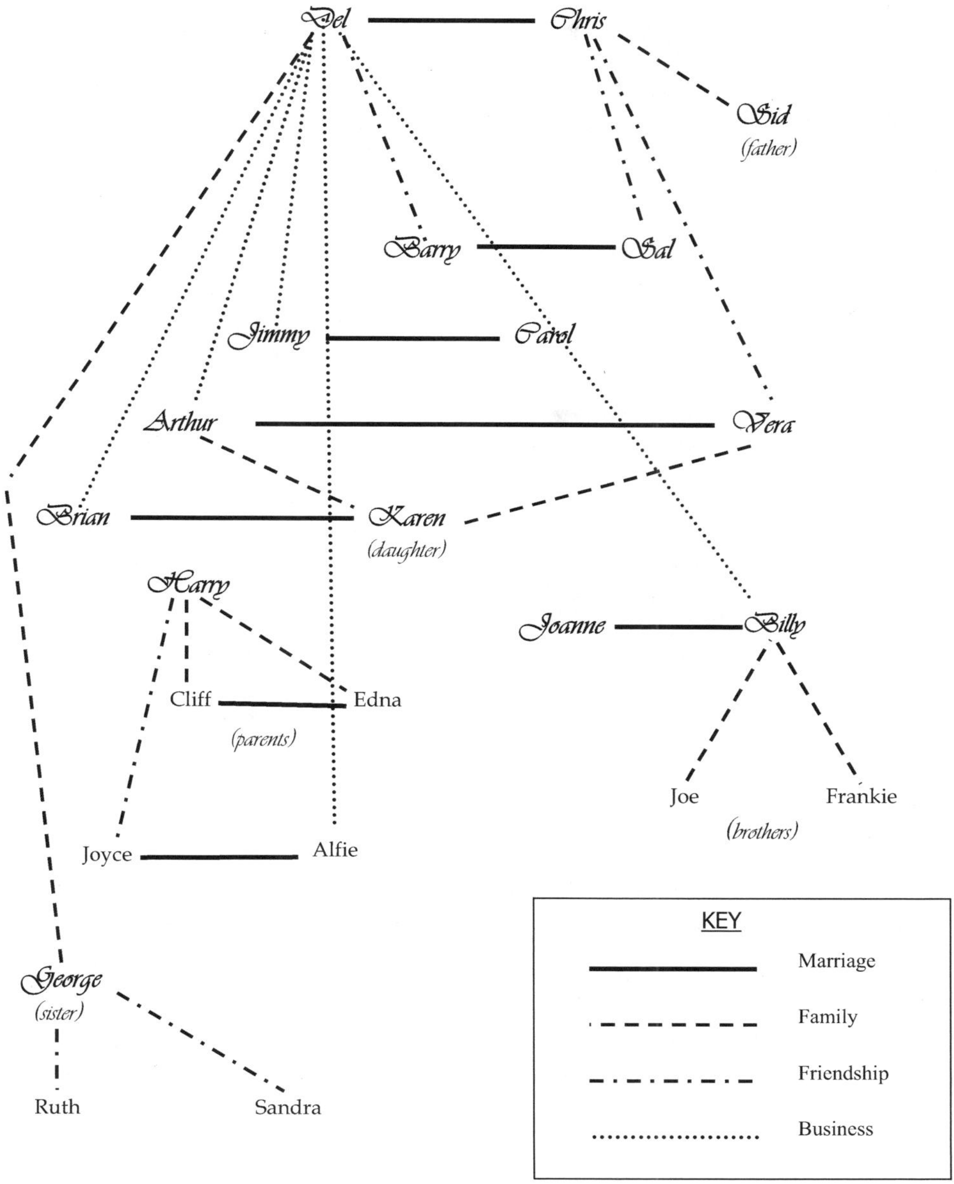

Acknowledgements to Janet Foster and Routledge ***After Foster, 1990***

Foster also produced 'portraits' of the pub 'regulars'; this is Harry, the barman:

> Harry was the regular evening barman. Although he had a full-time job, he worked in the Grafton most nights of the week and often spent his non-working

> hours there too – either way he drank heavily. Harry's role as barman was important because, like Del, he had contact with all the clientele. Captain of the men's darts team, Harry had a menacing appearance and looked like a stereotypical villain. He was never seen without several pieces of heavy gold jewellery which, he said, were acquired as a reward for doing some 'shady business'. He had served a long prison sentence for grievous bodily harm on his girlfriend. Harry claimed that he was a professional hit man for the South London underworld and was said to have several 'shooters' indoors.
>
> ***Foster, 1990:29***

Foster thought that some of this may have been bravado, and so decided to investigate Harry's 'boasts' a little further:

> I made what enquiries I could to confirm this information and the claim may not be as outlandish as it appears....There was enough support for these activities in Harry's case that I decided not to pursue taped interviews with him.
>
> ***Foster, 1990:51 note 2***

This defines the parameters of the group, but the subject of the (case) study was actually the group's rather dubious (criminal) activities/lifestyle. These activities (euphemistically referred to as 'doin' business') centred around one individual, in particular, 'Del' (see above). In order to find out how 'Del' and his associates view such 'activities', Foster conducted a series of interviews. These interviews were principally designed to answer three questions:

1) What was the difference between 'business' and crime?

2) What was their (Del and his associates) attitude towards crime?

3) What was their attitude to getting caught?

The answers to these questions can be found in the discourse elicited by the interviews (extracts from which are shown below):

What is the difference between 'business' and crime?

'Del' had interests in a scrap metal business (as well as the pub and a few other businesses), which meant that he was in a position to 'deal' in variety of goods. This is how he explains the difference between 'business' and 'crime':

> (N)o one could afford to sell a pair of shoes for five quid, but the guys in the pub can....In pubs there are always people that have got cars, they always want 'em repaired, or they want 'em to go from here to there every aspect of the business. If someone came up to me in the pub...[and] said, "I want a painting", automatically I'd think of Dave and I'd go, "I can arrange it." Therefore it now becomes business, that way I'd be lookin' at Dave sellin' his paintin' for £15, I'd

> be sellin' it for £20. I wouldn't think yeah Dave wants 15 quid, yeah I've got a pal who'll do that for 15 quid, that wouldn't enter me head – it's gotta be "yeah 20 pounds, he'll do the picture for yer." But I'd be on £5 of it.
>
> ***'Del' in Foster, 1990:36***

> If someone came up to me, say this guy with the shoes. As far as that's concerned I'll deal with anybody, because even if you're being screwed [set up by the police] you're only gonna say "yes" or "no" in a pub. Now if you're being screwed you're not doing wrong you're just doin' business, you know it's as simple as that. It's only when the guy goes, "I've got a pair of nicked shoes" and you go "Oh I'll have 'em", then you're in trouble you're no longer doin' business, you know what you're doin'. If the guy goes "Oh look I've got a pair of shoes for five quid, do you want 'em?" "Yeah, I'll have 'em", but as soon as he says "I've nicked 'em, they're five quid", then you're up against it, you know what you're about. It's a matter of how it comes across to me. You always know, no one would sell a pair of shoes for five quid, you know they're bent. The thing is unless you are told its only in your mind and in the court I don't think that stands up does it, your own opinion of things? it's gotta be like actual facts, so what I think doesn't really matter. I can only think oh the guy's nicked 'em but unless he tells me they're nicked, they're not nicked. So you work from that really, you always know, but it's a chance you take. At the end of the day, you've gotta argue with the police about it. I think out of things I've learned is how to answer the police.
>
> ***'Del' in Foster, 1990:38-9***

> I's say...[in the course of a week] you'd get eight to ten people come in... with bent gear...and ask yer whether you want somethink or the other.... You'd normally find at least twice a week, someone'll come in and go "I've got a mark four [Cortina] what parts do you want off it?", where they've gone and stolen the motor and got a nice set of wheels on it or wings or bonnet and they'll go and strip off what you want, and fetch it back to yer; or "I've got a box of tools, need any radios or batteries?"...Car radios, televisions come in quite often, stolen cars quite often, tools and things like that, a colossal amount of tools, CBs when they was the rage. You'll find that it's what's in the rage at the time, when videos first come out then you'd be flooded with videos. Radios, things like that, clothes come in ever such a lot, boxes of 'em. Gold comes in quite a lot, I mean necklaces, things like that on trays, and a hell of a lot of drink comes through.
>
> ***'Del' in Foster, 1990:39-40***

Clearly, 'Del' knows that the goods are stolen, but, because the person has not actually said "these goods are stolen" then they are not, technically, stolen. Even, "I've got a box of tools, need any radios or batteries?" (rather than the more usual "I have a battery or radio to sell") is not enough to make such goods 'stolen'. This shows how 'Del' and his associates almost convince themselves that they are not dealing in stolen goods. This brings us on to the second question:

What is their (Del and his associates) attitude towards crime?

This is how Del's wife sees it:

> If someone came up and offered me a suede coat for £20 and it fitted, I wouldn't ask no questions, I'd take it. But then most people are like that, aren't they.
>
> ***'Chris' in Foster, 1990: 43***

Sentiments, which are echoed by one of her friends:

> I think we all do it sooner or later. We all know they're stolen although you might not have stole 'em, we all know. I've had 'em, they've been cheap, like if someone offers you a cheap washing machine and you know it was stolen, well it must be stolen for a brand new washing machine say for a hundred pounds or a video, you're not gonna ask no questions, you're gonna buy it because it's cheap and I think that's how everyone thinks. Even me mum and dad would have it.
>
> ***'Joanne' in Foster, 1990: 43-4***

> It's wrong but then I do it. The way I look at it, I didn't steal it so I'm not worried about it. I suppose it would be a different story if I got found out and the police done me for accepting stolen property.
>
> ***'Joanne' in Foster, 1990: 46***

Notice, here, that the women are placing themselves in the role of 'end user', rather than dealer (or even thief). They play down the 'criminal' aspect of the activity, by arguing that 'receiving' is not as bad as stealing. They also attempt to make this behaviour appear 'reasonable/understandable', by arguing that we would all buy cheap goods, if we had the opportunity. (Even if one suspects that the goods are stolen.) This 'explanation' is consistent with the distinction drawn by 'Del' (between 'business' and 'crime').

With regard to *who* they steal from, 'Del' and his associates see themselves as only stealing from the 'rich'. Here again, 'Del' appears to be having difficulty reconciling what he knows with what he wants to believe:

> I don't particularly like the idea of people saying I don't steal off me own; I only steal off the rich. I could be the classic example seven years ago I didn't have two bob, people 'd steal off me now. I'm still one of 'em. In fact I'm as much one of 'em now as I've ever been. So I don't believe that I won't steal off me own cos everybody's yer own, the only ones that aren't yer own are the ones that are born with silver spoons.
>
> ***'Del' in Foster, 1990: 48***

Having accepted, to some degree, that what he is doing is criminal, there is only one question remaining:

What is their attitude to getting caught?

This is Del's reply:

> You never think about it before you do it. You don't think I might not get away with it, it is only when they've got yer in their hand you think they've got me.... You don't think police while you're thinkin' steal. The two don't go. I think I've always had this thing where I've known whether I'm gonna get caught of not gonna get caught. You know when you're doin' it, you know full well that you're gonna go in there and you're gonna get caught and sod me you still go and do it! There's somethink tellin' yer you're gonna get caught this time. You could have been in the same place five, ten times and know full well you're gonna walk in and out, and one day you walk in and think it's gonna happen today and it would.
>
> ***'Del' in Foster, 1990: 49***

These extracts show that 'Del' and his associates have developed (although the credit perhaps belongs elsewhere) a number of devices, which allow them to put the wrongfulness of these activities out of their mind – it is almost a belief system! This shows that the group is not only held together by family and 'business' ties, but (perhaps necessarily) the same outlook on life.

Returning to the methodology of the design, we can see that the group constitutes a 'unit' (making it suitable material for a case study), and (by definition) that the study provides an account of that 'unit', *as a whole*. The investigation identifies this 'unit' in terms of its members, their relationship to each other, and their beliefs, each of which contributes to our understanding of the 'unit', as a whole. It also locates, and distinguishes the 'unit' from its physical and social setting. The extracts presented here, demonstrate why (in my opinion) this is a good example of case study, although the level of detail, used in the (actual) investigation, makes it difficult to see the subject *as a whole*.

Section Conclusion

With any case *study*, the researcher must first define the 'case' (the subject of the investigation). This means that the 'case' must have 'boundaries', which distinguishes it as such. With the first example, the individual is distinguishable by virtue of his experiences, and these experiences are bounded by two events – the operation and the individual's demise. The 'case' is not the individual, but the individual's perspective. With the second example, the public house is distinguishable by virtue of its physical location, and 'Del' and his associates are distinguishable by virtue of their lifestyle. The 'case' is not the public house, but the lifestyle of certain patrons. Therefore, a 'case' does not necessarily have to be a *formally* defined/bounded 'unit' (such as an organisation or an officially recognised group). The case study is also the justification for its own design: the subject ('unit') justifies studying it as a 'case', which justifies the ('case study') design. In other words, the subject itself ('unit') suggests that something can be leaned from studying it *as a 'unit'*. In both of the above examples, there is something *to be learned* from studying them as 'cases', which makes it *appropriate* to study them as cases, rather than as an assembly of concepts. The first looks at how a particular individual makes sense of the world; and the second looks at a particular social network. Both of these are *holistic* in nature, making it not only rewarding, but appropriate to study them as 'cases'. In this way, the reason for studying the subject, effectively, becomes the research question (as well as the justification for the design and the choice of subject).

Clearly, the methods chosen to investigate the subject, will need to be the ones which enable the researcher to answer the research question (to arrive at a holistic understanding of the subject). With the first example, the subject was the behaviour of this individual, which made (naïve) 'experiments' and 'observations' the most appropriate methods to use. With the second example, the subject was the lifestyle of these individuals, which made individual interviews the most appropriate method to use. Both studies also made use of anecdotal observations and (as appropriate) information derived from other methods/ sources. (Whatever was necessary to provide an understanding of the 'unit', *as a whole*.) While they may have been designed to answer a particular research question (a 'case study' of the subject), it is important not to ignore their heuristic value (see Taylor, 2007). A piece of research not only contributes to knowledge by answering its own research question, but also by suggesting other research questions. The second investigation is good example: not only does it consist of a 'case study', but in-depth investigations of a number of topics (such as how the women feel about their lifestyle) arising out of that (case) study.

Published/Public Information - Written

In this section, we will look at how published/public information can be used to answer different types of research questions. Unlike the documents and sources discussed in previous chapters, those discussed here reside (predominantly) in the *public* domain (as archives, publications, or broadcast material). As with most of the descriptors used in this typology, the term 'public' is used very loosely here. A great many 'public' archives, record offices and libraries are described as 'public, even though they may require those 'members of the public' to make appointments, apply for 'membership', or justify their need to use the material (which usually means for research purposes). In other words, even in the case of public archives/records etc., 'public' does not (necessarily) mean that 'just anyone' can go in, and browse around, as they please. The same is true of archives, held at research institutions, which tend to be 'open', only, to 'members of the academic community'. It is unlikely that an archive will turn away a 'serious' researcher, just because they do not have any formal qualifications. (One does not need formal qualifications to be a member of the academic community.) By definition, these archives are only likely to be of use to people who have a 'serious' interest in the subject, anyway. The word 'public' is simply being used, here, to distinguish between those records/documents, which are held *for* 'public' scrutiny, as opposed to those in the 'private' sector that are *not* intended for public scrutiny. (In this respect, legal records are, perhaps, more 'public' than 'private'.) It is also the case that Ph.D. (Doctor of Philosophy Degree) theses are deemed to be 'public documents' – which means that anyone can ask (the awarding institution) to see the thesis of any Ph.D., awarded by any institution, in this country (England). (They are, however, at liberty to make a charge for this service.)

The category *'Published/Public – Written'*, clearly refers to 'published' material, but, here again, the term 'published' is not *quite* as straight-forward as it first appears. When people think of 'a publication', they tend to think of a *book*, published by one of the major publishing houses. This is the traditional conception of a 'publication', but a great deal of published material does not conform to this 'traditional' conception (of a publication). For example, a great deal of research is published in academic journals. Academic journals are themed (according to the area of academic interest) periodicals, used in the dissemination of academic research (see Taylor, 2007). Although academic journals tend to cater for a very specialist audience, they tend to be published by the major publishing houses, and/or professional bodies. Another type of 'minority interest' publication is 'the newsletter' (as a type of publication). Newsletters are 'published' by charities, and interest groups, *for the benefit of their members*. In other words, the membership is also the intended readership (which also gives a clue to the content of the publication). Newsletters may not be as professionally produced, or have as large a print-run, but they are nonetheless 'published'. (The *'Collins English Dictionary'* defines the word 'published' as "make generally known". Here, the word 'general' is very much like the word 'public' (in that it is open to a certain amount of interpretation), which is why newsletters can be regarded as (and are, for the purposes of this typology) a publication.) For the purposes of this typology, all material produced by the mass media (including the Internet) is classed as 'verbal' *('Published/Public – Verbal')*. (The two obvious anomalies here are, of course, the press and the Internet.) As in chapter two, the written/ verbal distinction is somewhat arbitrary, here, in that 'the mass media' represents an extension of the above ('written') sources, rather than a completely different class of material. To reiterate the point made earlier (in chapter one), this typology is simply intended to provide a rationale for the order of the topics, it is not designed to be a system of classification *per se*.

Unlike individuals, or private companies, who may record and/or disseminate information as a 'by-product' of their daily existence (even if it is writing their autobiography, or selling a list of clients to an advertiser), public/published sources exist, solely, *for the purpose* of collecting and/or disseminating information. In other words, they are *'in the business'* of collecting and/or disseminating information. If the source (of the information) is an individual or private company, then the researcher is expected to evaluate that information, in order to ensure that it comes up to the standards of the academic community. However, if the source (or the information) is an organisation, which has been set up for the purposes of providing information on a particular subject (i.e., a 'professional'), then it seems reasonable to expect the organisation to evaluate (and maintain to a high standard) the quality of the information it provides. This discussion will, therefore, focus on the 'quality controls', which these sources/organisations are likely to have in place, and the reasons why a researcher *can* have confidence in the quality of the information. The topic area is divided into three sub-sections: archives, publications (written) and the media (verbal). The section on archives will look at the function and logistics of an archive, using the *'The British Cartoon Archive'*, at the University of Kent at Canterbury, as an example. It will also look at what happens when an archive fails to maintain the quality of its information. The example chosen to illustrate this scenario is the *'Eugenics Record Office (ERO)'*, which was located at Cold Spring Harbour, New York, between 1910 and 1939. The section, on publications, will look at why one can have confidence in the credibility of certain publications/authors (academic publications, government publications, and charity/interest group newsletters). This theme is continued into the section on 'media', with a critical discussion of news media research.

Published/Public Written	**Archives and Publications**
Published/Public Verbal	**The Media**

Archives

Archives are collections of material which have been set up for the purposes of providing a resource on a particular subject. For example, *The British Cartoon Archive* (formerly *The Centre for the Study of Cartoons and Caricature)* at the University of Kent, at Canterbury, was set up:

> (T)o collect, conserve and make available the originals of, primarily, 20th-century British newspaper cartoons – to pursue and promote a wide-range of interests and research in cartoons.
>
> ***Information supplied by The Archive, 2001***

Gathering the relevant material is, therefore, a large part of the archive's work (as it is with all archives). (In this respect, it is a 'typical' archive: even if its subject matter is not.) Thus far (in the book), we have only considered 'information gathering', as a task directed towards answering a particular research question. Archives gather information in order to act as a resource on the subject. This means that the information has to be organised in such a way that it can be configured to answer any conceivable (and perhaps inconceivable) (research) question, asked from any conceivable perspective. For example:

> Users of the Cartoon Study Centre include authors, newspapers, TV companies, publishers, enthusiasts and GCSE examiners.
>
> ***Information supplied by The Archive, 2001***

This is quite easy to demonstrate: reproduced below is an item supplied by *The Archive.* Before looking at how *The Archive* classifies this item, think about how *you* would do it.

"Couple of gins and tonic, a peck on the cheek with the office typist and to you it's an ORGY"

Acknowledgements to Langdon and *The Sunday Mirror*

The Archive uses no fewer than ten descriptors to classify the content of this item. This is in addition to the format, author, date, and source publication:

```
27½ x 20 cm    Ink and blue wash

SOCIAL EVENTS : parties
INTERIORS : domestic - halls
MARRIAGE AND SEXUAL RELATIONS : domestic relations
DRINKING : wines and spirits
WORKERS : typists
MARRIAGE AND SEXUAL RELATIONS : sexual relations - kissing
SOCIAL EVENTS : orgies
              : parties - balloons
```

Acknowledgements to Jane Newton, and *The British Cartoon Archive*, University of Kent

This is a very similar system (of classification) to that used in **databases**. In a database, items are classified by a number of descriptors. These descriptors are then used to access the item. Searching for material, on a particular subject, means searching for items classified under that (and/or a similar/related) descriptor. (Similar to the 'key word', search facility, on most library catalogues.) It is this, which enables an archive to perform its function (as a resource for researchers). An archive's ability to perform its function (as a resource for researchers), therefore, depends on the quality (and quantity) of the material it holds, and the ease with which (relevant sub-sets of) that material can be accessed. If an archive fails on one, or both, points, then its utility must be called into question. This is precisely what happened to The Eugenics Record Office (ERO) at Cold Spring Harbour, New York (Allen, 1986). Let us now look at the details of this case.

The Eugenics Record Office (ERO), at Cold Spring Harbour (1910-1939)

In order to appreciate the role and function of this archive, it is necessary to look at the period of social history in which it existed. The end of the nineteenth century saw a number of pioneering studies in the (recently discovered) field of genetics: colour blindness, A-B-O blood groupings and several metabolic diseases were all found to follow patterns of Mendelian genetics[4]. In 1904, Charles Davenport was able to persuade the Carnegie Institution of Washington to provide the funding for a centre into the study of Mendelian genetics – *The Station for the Experimental Study of Evolution (SEE).* He was convinced that the theory could rival Darwinism (as an explanation for evolution). Encouraged by the centre's success (into plant and animal genetics), Davenport proposed a sister organisation (*The Eugenics Record Office (ERO)*), which would apply the idea of Mendelian genetics to *human* 'evolution' (Allen, 1986). At this point, the history of the ERO starts to get a little vague – it is not clear why the Carnegie

4. Genes can be either dominant or recessive – those carrying a deformity are usually recessive – any combination of normal-normal or normal-deformity (gene) will produce a 'normal' plant/creature/individual – the deformity-deformity combination will produce a 'deformed' plant/creature/individual – off-spring inherit one gene from each parent – parents may have either the normal gene or the deformity gene – making the probability of a 'deformed' off-spring: one-in-four – according to Mendelian genetics.

Institution did not want to fund the ERO. This did not deter Davenport, who used his connections (Allen, 1986) to obtain backing from the wife of the late railroad magnate, Mrs Mary Harriman. It was also around this time that the first Eugenics Committees were being set up. These eugenical societies attracted the interests of prominent evolutionists; they also attracted the interests of entrepreneurs, such as Vernon Kellogg and Alexander Graham Bell (Allen, 1986). This begs the question: *'why would the wife of a railroad owner, the inventor of the telephone, and a cereal manufacturer, be interested in biological research?'* One can perhaps see why Vernon Kellogg would be interested in such research. But Mary Harriman and Alexander Graham Bell? One 'explanation' can be found in the eulogy, given by Davenport, at Mrs. Harriman's funeral:

> As she often said the fact that she was brought up among well bred race horses helped her to appreciate the importance of a project to study hereditary and good breeding in man.
>
> ***Davenport, eulogy "Mrs Harriman", quoted in Allan, 1986:236***

While this is *one* 'explanation', it is not really good enough to explain why the Harriman family contributed an estimated (Allen, 1986), $395,000 (1910-20s) to the project. Allen offers us a different explanation – the fear of crime (and social degeneration). At the time, a lot of Eastern Europeans went to America to 'seek their fortune'. It perhaps goes without saying that many failed, and 'had to' turn to crime in order to support themselves (and their families). This 'failure' was seen as a product of a genetically *inferior* temperament. Davenport's work (at the ERO) held out the promise of finding 'evidence' that would support this theory. This 'evidence' would allow the authorities to instigate legislation, which would allow them to exclude individuals from 'races' with a predisposition to failing.

At the time (of its conception), there was no reason to think that the ERO would be a disgrace – Davenport was a respected academic; the SEE was a respected institution, and the way in which the ERO was set up was not 'unprofessional'. Its (the ERO's) aims and functions were clearly and professionally laid out (on the inside back cover of each ERO Bulletin), and these were:

> 1) To serve eugenical interests in the capacity of a repository and clearing house.
> 2) To build up an analytical index of inborn traits of American families.
> 3) To train field workers to gather data of eugenical import.
> 4) To maintain a field force actually engaged in gathering such data.
> 5) To co-operate with other institutions and with persons concerned with eugenical study.
> 6) To investigate the manner of the inheritance of specific human traits.
> 7) To investigate other eugenical factors, such as (a) mate selection, (b) differential fecundity, (c) differential survival, and (d) differential migration.
> 8) To advise concerning the eugenical fitness of proposed marriages
> 9) To publish results of research.
>
> ***ERO Bulletin, 1916, inside back cover***

Even so, this list of aims and functions shows that, from the start, the ERO's 'loyalties' were divided. The ERO had to satisfy the needs of both academics (as a resource for researchers) and patrons (providing evidence for racial segregation). (This would ultimately be its downfall.) This can easily be demonstrated by separating this list of aims and functions into those, which an institution of this type would, ordinarily, be expected to have (1,3,4,5 and 9), and those, which are related to the 'application' of eugenics (2,6,7, and 8). The former (group of functions) established the ERO's academic credentials (as a research resource), while the latter satisfied its patrons.

In their paper, *'How to Make a Eugenical Family Study'* (ERO Bulletin, 1915) Davenport and Laughlin list eight classes of traits (from August Hoch and George Amsden's *'The Family Distribution of Personal Traits'*), divided into 150 item questions. (We would now call these 'diagnostic criteria'.) They also list the average weight men and women (based on height and age), and 68 'anthropometric measures' (physical characteristics), "usually taken in gymnasia" (ibid. 16-17). Davenport and Laughlin did not use (all of) these measures. Instead, they chose to use their own: *'Individual Analysis Card'*. These *'Individual Analysis Cards'* had just 60 items, and covered everything from the person's name and address, to their temperament and personal appearance. Although there is no record of this instrument ever being 'tested' (for effectiveness, reliability, validity, etc.), the advice, which Davenport gave to prospective users, shows that he wanted these schedules to be administered competently and consistently:

> Referring again to the Individual Analysis Card, it is respectfully urged that the questions should be carefully examined before an attempt is made to fill them out. In order to become acquainted with the schedule fill it out first for yourself and next for those of your family whom you know most intimately. Distinguish between traits that are constant and those that are periodic; also between tastes and abilities, i.e. one may be fond of hearing music but incapable of rendering it. The fact that a card cannot be filled out completely should not discourage one from attempting the analysis; even a few facts concerning certain individuals may be of great importance.
>
> ***Davenport and Laughlin, 1915:24***

These 'Individual Analysis Cards' were used to produce family 'pedigree charts', which were, essentially, 'maps' of (hereditary) *family* characteristics. The 'pedigree charts' for the Hutchinson family (showing their 'Natural Musical Ability') and the Dack family (who were not quite so gifted) are reproduced on the following pages. (Squares represent males; circles represent females; horizontal lines represent a generation; and shaded areas (or other notation) represent the presence of an identified trait.)

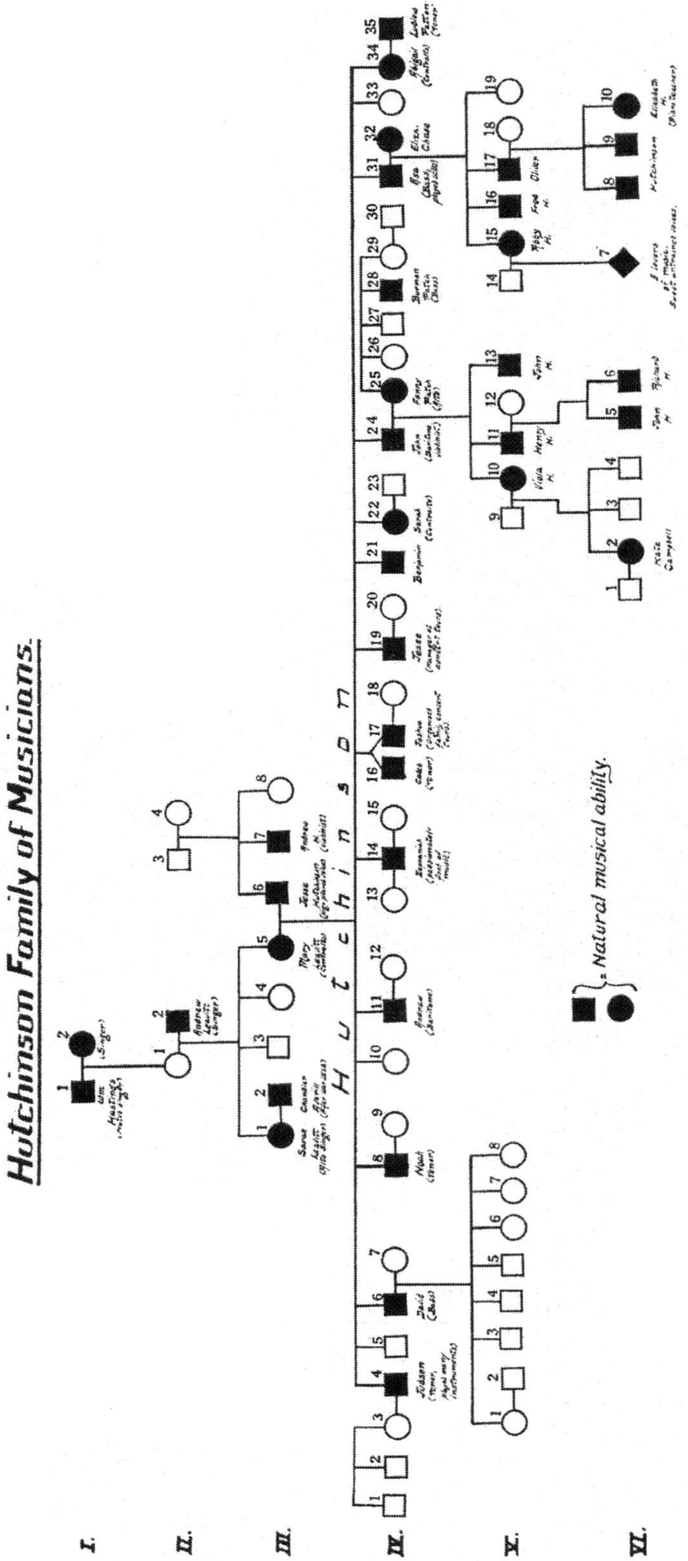

FIGURE D.

Acknowledgements to Charles Davenport and Harry Laughlin, The Eugenics Record Office, Clare Bunce and The Cold Spring Harbor Laboratory Archive.

CHART OF THE DACK FAMILY

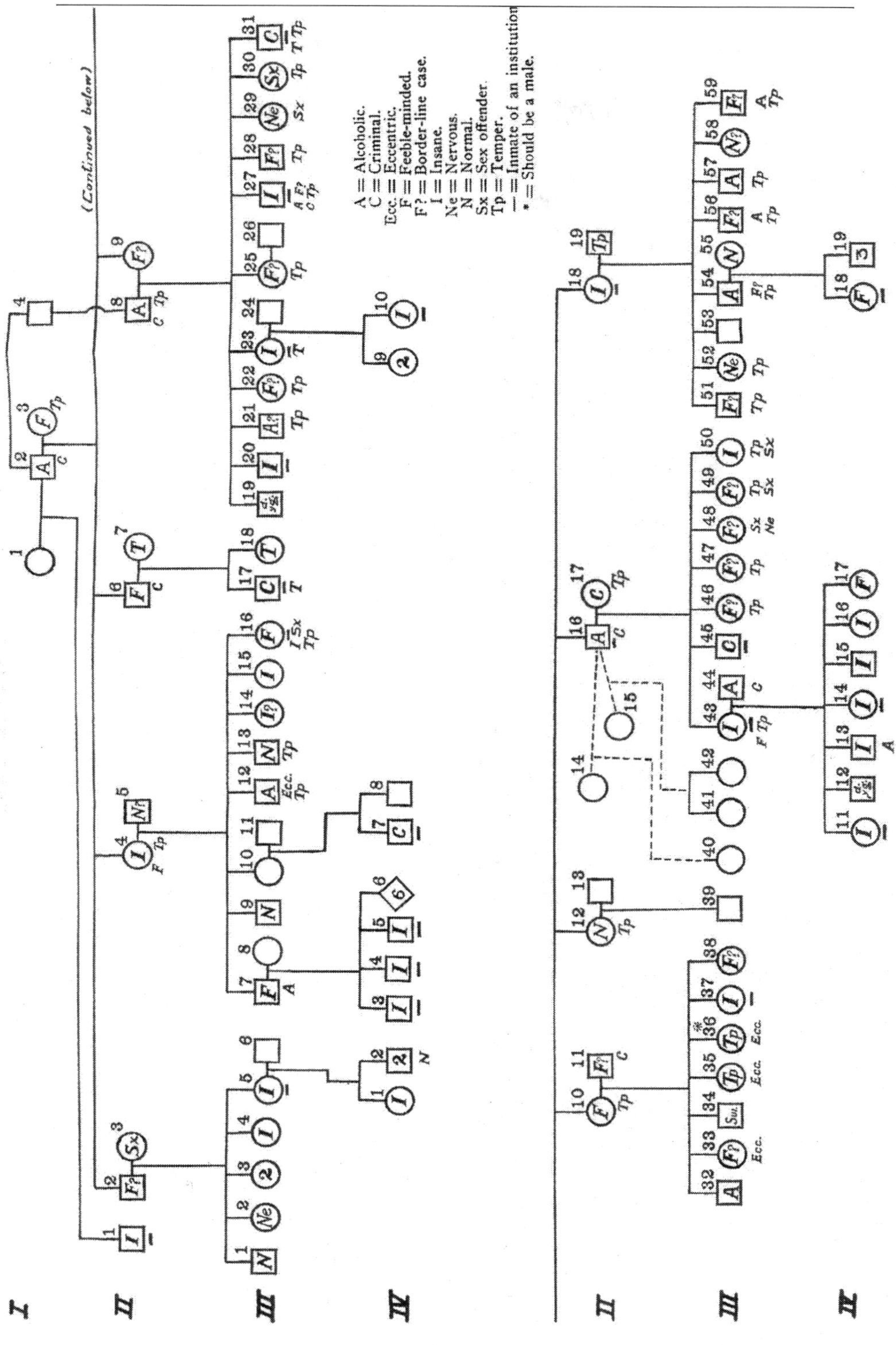

Acknowledgements to Anna Finlayson, The Eugenics Record Office, Clare Bunce and The Cold Spring Harbor Laboratory Archive

The ERO actively encouraged people to contribute to its collection of 'family pedigrees' – the inside, back cover of the ERO bulletin reads "These schedules [*'Individual Analysis Cards'*] are sent free in duplicate to such persons as will take to fill them out and after retaining one copy for their own use, will file the other with the Eugenics Record Office." (*ERO Bulletin*, 1916:inside, back cover). These 'family pedigrees' became the main-stay of the ERO. They did not, however, attempt to determine the (genetic of social) origin of the 'trait'. Davenport believed that this was something which could be investigated at a later date:

> It is not desirable at this time [1915] to attempt to differentiate between inheritable and non-inheritable traits. Many physical, mental, and moral traits have been proved to have a hereditary basis, and it seems probable that in practically all there is an hereditary factor of more or less importance. However, precise knowledge of the hereditary factor is often lacking and, among other things, the recording of family data will eventually afford the means of studying this hereditary factor. The practical rule is to record all of certain classes of facts concerning as many members as possible.
>
> ***Davenport and Laughlin, 1915:3***

Anna Finlayson's (1916) study of the Dack family (from which the second chart is taken) - *'The Dack Family: A Study in Hereditary Lack of Emotional Control'*, was considered (by Davenport, in the study's preface) to be a particularly good example of such research:

> The family history that Mrs. Finlayson has recorded in the following pages is a good example of the results that accrue to a careful investigation made by a eugenics field worker – results that can be secured in no other way.
>
> ***Davenport, 1916:iii***

In addition to the 'pedigree chart', Mrs. Finlayson's study also included descriptions (characterisations) of family members. These descriptions were then used to sum up the Dack family and its 'contribution' to society. While Davenport may have seen it as a 'careful investigation', the report's descriptions were heavily influenced by the cultural stereotypes and political rhetoric of the day. This can clearly be seen in the report's conclusions:

> *The Dack family:*
>
> In making an analysis of [153 members of the family, and] the various peculiarities found in the family, we find twenty-five cases of insanity, only two of which were senile psychoses; twenty individuals are described as being lazy or shiftless; thirty-nine are below average intelligence; thirty-four are described as ugly, quarrelsome and bad tempered....Fifteen other individuals are described as merely quick tempered. Thirty are mentioned as being alcoholic...In seven instances there was a history of thieving without jail sentences; in six others there was incarceration for breaking the law...eleven other individuals have a history of dishonesty, making, in all, a total of twenty–four dishonest individuals

with thieving tendencies....We find mention made of sexual irregularity in twenty-seven individuals, practically all females. In some instances merely illicit relationships previous to marriage is recorded, while in others cases of prostitutes.
One trait which the reader has probably already noted is the tendency shown by members of the family to desert the consort for a short period of time. This tendency is mentioned in eighteen instances. One woman in the family has kept up for years this practice of separating from her husband and later coming back to him. Nine individuals are described as neurotic or excitable, and six others as eccentric.
(A)ll individuals under twenty have been excluded. Many of the family who are below that age have already shown themselves to be defective, however; one child is an idiot, four others are recognized imbeciles, one of whom is in an institution, two are epileptics, several young girls immoral, seven markedly neurotic, and a large number show evidences of some degree of mental deficiency.

Finlayson, 1916:40-41

The causes:

In considering which branches of the family are the most defective, the descendants of Jane and Curtis are immediately singled out. The reason why we find more degeneracy among Jane's descendants then in the rest of the family is found, probably, in the fact that she married her cousin. The reason in the case of Curtis is not quite so apparent. He married, it is true, a defective woman from defective stock, but his sister Isabelle contracted an equally bad marriage with rather better results. It is in the families of Jane and Curtis that we find the most pronounced cases of lack of emotional control. In both cases this trait was inherited from both sides of the family....This cousin marriage seems to bring to light in a rather ghastly fashion the worst of the family traits.

Rebecca Dack married a thrifty German and among her children and grandchildren we can trace, to a certain extent, the blending of the two families. The slow, plodding German nature seems to have counteracted the quick Irish temper of the Dacks, for comparatively few instances of marked aggressive temper are found among her children, and the largest number of steady, capable children are found among her offspring....The evidence presented seems to corroborate Dr. Davenport's theory that a quick temper is inherited as a Mendelian dominant, for the trait does not skip a generation.

Finlayson, 1916:41-44

The family's contribution to society:

Almost every individual in the family is endowed with enough practical ability to make a living. But as a family hey are lacking in powers of self-control, in ambition, and in higher reasoning ability. Their lack of self-control, shown by violent outbursts of temper, has already been considered. Closely related to this

> is a lack of stability shown especially in the matter of marriage relationship and residence....The family, of course, has been an economic loss to society. Ten of them have been or are patients at the Warren State Hospital. Their total length of residence is seventy-four years and two months. Figuring the cost to the state and county of these patients, we find that, until August 31, 1915, these patients cost the taxpayers of the state $16,354....An attempt was made to estimate roughly the expense to the state of the individuals who have been in other institutions – county homes, other hospitals for the insane, penitentiaries etc. Allowing $200 per year per person, this would make about $12,000, which would make a total of $28,354 spent in a period of forty years. Looking at the matter from an economic standpoint only, this figure would not be so alarming if the members of the family who had not been in institutions were productive and a source of wealth to society. Some few of them have been, but not the majority. Furthermore, many of them who have not been in institutions have been a cause of economic loss because of their dishonesty and theft. No attempt has been made to estimate the cost of occasional jail sentences or the money that has been paid in poor relief to the family...
>
> The family has also been a decided detriment to society from a moral standpoint, and this loss, which can not be estimated in dollars and cents, is more important than the economic loss.
>
> ***Finlayson, 1916:43-44***

[All in all, a family which society would be better off without!]

While, to us, this account may seem 'impressionistic' and value-laden, to Davenport's philanthropists, it was exactly the kind of thing they wanted to hear. This investigation was 'evidence' that a degenerate, immigrant family was incapable of being anything else. It was the kind of 'evidence' on which legislation could be (and was) based (Allen, 1986).

For reasons which are not entirely clear, the Harriman family's philanthropy ceased in 1917, and the Carnegie Institution took over the financing of the ERO (Allen, 1986). While the aims and objectives of the ERO shows that Davenport intended it to be able to accommodate the interests of both philanthropists and academics, its ability to satisfy the latter had not been tested (up to that point). In 1929, the Carnegie Institution set up a committee to investigate the viability of the ERO. The ERO's survival depended on Davenport's ability to convince his new (academic) philanthropists that the ERO deserved their (continued) support. Davenport resigned from the ERO (leaving Laughlin in charge) just a few months before the Advisory Committee (of the Carnegie Institution) was due to make its inspection. When it did, its conclusions were as follows:

> The records, upon which so much effort and money has been expended, have to date been extremely little use, to judge by the number of publications based on them. Thus the [Eugenics Record] Office appears to be accumulating large amounts of material, and devoting a disproportionately great amount of time

> and money to a futile system for indexing it, without certainty, or even good probability, that it will ever be of value.
>
> ***Report of the Advisory Committee of the Eugenics Record Office: pp 2-3, in Allen, 1986:251***

Specifically:

> They found the ERO's total collection of records "unsatisfactory for the study of human genetics"...traits such as "personality" or "character," lacked precise definition or quantitative methods of measurements...traits, such as "sense of humor," "self respect," "loyalty," and "holding a grudge," could seldom be known to anyone outside an individual's close friends and associates...to get an honest recording of such traits is virtually impossible....and even more objectively measurable characteristics, such as hair form, eye colour, or degree of tooth decay, become "relatively worthless items of data for genetic study" when recorded by an untrained observer.
>
> ***Allen, 1986:251-2***

The ERO was finally closed down (by the Carnegie Institution) on 31st December 1939 (Allen,1986).

Methodologically speaking, Davenport did a lot of things *right*: the ERO had a good academic foundation; it had clear professional objectives; it sought to measure constructs professionally and methodically; and its instruments were administered along clear, professional guidelines. In spite of its promising start, Davenport allowed the ERO to be ruled (too much) by 'politics'. Consequently, insufficient attention was paid to defining and/or measuring constructs, and the 'investigations' tended to cater for his patrons, rather than the academic community. These factors conspired to produce 'findings', which were more 'political' than 'scientific' (and, therefore, of little interest to academics). It had yielded few academic publications, and was, therefore, assumed (by the Carnegie Institution) to be of little use to academics. Although, the ERO had not failed by its own standards, it *had* failed, by the standards of the academic community. This, rather conveniently, brings us on to the subject of publications (and academic publications).

Publications

The Publication as/by an 'Authority'

Very often, the only way to judge to quality of the information is to ask whether or not the person can speak from a position of authority (on the subject). This does not mean that they are in a position recognised as giving them that authority, merely that they are in a position to know about the subject. Publication carries with it the implication that it has been written by an 'authority' on the subject. This begs the question: *'who is an authority?'* The term 'authority' is so vague that almost everyone can claim to be an 'authority' on *something*. It is often the case that an 'authority' will claim to have 'expertise' on the subject, which begs the question: *'who is an expert?'* This simply seems to be replacing one ill defined concept with another. While it takes no formal qualifications to be an 'expert', the term does, at least, imply an extensive knowledge of the subject. This raises the question: *'how much knowledge is required to be an expert?'* Estimates vary, but on average it takes around 4,000 hours, or 10 years, to become an expert (Smyth, Morris, Levy and Ellis, 1988:309). (You may, however, be able to think of people who consider themselves 'experts', with a lot less knowledge than this.) Straight away, one can begin to see the problem here (and it is one of 'quality control'). How can anyone be sure that a publication has been written by someone who can write *'with authority'* on the subject? This means looking for reasons to have confidence in the quality of information (included in the publication). In the following discussion, we will look at how information is sanctioned/validated, in three types of publications: academic, government/official, and interest group/charity newsletters. The discussion will be neither exhaustive not exclusive, it will simply look at three different types of publication, in terms of how confident we can be in the information they publish.

Academic Publications

Academic journals have a system (of quality control) known as **anonymous peer review**, authors submit their article/ 'contribution'/ 'submission' (to the journal), with their name and contact details (usually their affiliation: the organisation or university department where they are currently working) printed on a *detachable* cover sheet. The name and details of the author do not, therefore, appear on the article itself. This ensures that the article can be read *anonymously* (without knowing (or having any clues to) the identity of the author). The editor (of the journal) then sends the manuscript (minus the cover sheet) to two referees (academics with a knowledge of the subject), who then review the article and send their opinions (regarding the quality and/or publishability of the work) back to the editor *anonymously*. The editor then decides (on the *basis* of the referees' opinions) whether or not to publish it. This decision is then communicated to the author. Although this may seem like a *'final judgement'*, 'rejection' of an article (by the referees and/or editor) is not as damning as it first appears. I have known of cases where an author has appealed (to the editor) against a referee's opinion, and the article has subsequently been published. I have also known of cases where an article has been accepted by one journal, but not another. The reason for this is that some articles will appeal to some journals more than others. It is, therefore, important (when submitting an article to a journal) to make sure that one's article matches the interests of the journal. By way of an illustration, here is the editorial policy of the journal *Media, Culture and Society*:

> *Media, Culture and Society* provides an important international forum for the presentation of research and discussion across the whole field of cultural practice. To date, the main focus has been on the mass media (television, radio, and the press) within their political, cultural and historical contexts. But a developing additional focus is a critical engagement with the concept of the Information Society and with those issues raised by the convergence of the mass media with systems of cultural production and diffusion based upon telecommunications and computing. We are also increasingly interested in the social role of various categories of intellectual in the full range of cultural practice. The relationship of all these systems and concerns to literature, the visual and performing arts, photography, publishing and to more general artistic and cultural practices is of

central reference. The journal seeks to relate academic work to contemporary cultural practice, particularly with regard to mass media and communication policy. As we are an interdisciplinary journal, contributions from people in any relevant areas are welcome. The journal is also committed to making available the work of foreign scholars and practitioners (via translation and specifically commissioned articles) as well as of American and other English-speaking countries. Contributions to the journal are externally refereed.

'Notes for Authors', 'Media, Culture and Society'

This tells us that the journal is prepared to accept submissions written from any perspective (discipline), but the work *must* consider the social and/or political context of the subject. Therefore, an article which simply describes the content of a medium (without relating this to its wider social/political context), is unlikely to be accepted by the journal. This ('advice note') also tells us what kind of articles we are likely to find in the journal, as well as the fact that they have been 'externally refereed'.

The strength of the procedure (anonymous peer review) lies in its ability to maintain the author's (and, to a lesser extent, the referees') anonymity, thus ensuring that the referees will be *impartial*. On the question of anonymity, some academics have pointed out that a knowledgeable referee may be able to guess the identity of an author (from the subject and the literature cited), thus undermining the impartiality of the procedure. While this is always a possibility, it is difficult to see how a referee can have knowledge of the subject without knowing who is working in what area! Even if a referee is able to guess the identity of an author, there is nothing to say that it *will* influence their opinion of the article. In spite of such 'weaknesses', anonymous peer review is still one of the *best* quality controls, used in any form of publication (which brings us on to these *other forms* of publication).

Government Publications

Although government publications (official reports, statistics, etc.) are 'public' in the sense that they are accessible to the public, it is questionable whether they are 'public' in the sense that they are 'for use by the public' (as with an archive). While it stands to reason that Government research must be of a reasonably high standard (given that it is the basis of policy decisions), the 'findings' can be difficult to interpret without knowing the precise (research) question which the research was designed to answer. Likewise, 'published' and 'for public reading' does not necessarily mean that their *full* meaning can be understood by 'lay' individuals. Arguably, 'government publications' are more akin to the documents use by private companies (in the running of their business) than the material held in public archives. A good example is the (on-going) British Crime Survey. The British Crime Survey shows the Government where its resources can be used to best effect (in order to combat crime). This is the case with any 'social problem'. If something seems to be causing public concern, then the Government will (commission someone to) investigate the problem (and perhaps see how it can be addressed). In the same way, if a particular (public) policy, department, or institution seems to be failing, then the Government will commission an investigation into 'the problem'. Although the findings of these investigations are always made available to the 'public' (published by Her Majesty's Stationary Office: HMSO), one must not lose sight of the fact that the research has been commissioned by the Government, in order for it to fulfil its function (as a government). A number of academics have been instrumental in highlighting the 'political' function which such research *could* serve. While there is no doubting the validity of such observations, it is, perhaps, questionable how much of a 'secret' this really is (or how profound it is to say that government research reflects the politics of those in power).

For this reason, academics tend not to take official (government) statistics at face value (as a source of information *per se*), but use them (the subject matter and design) as indicators of contemporary political thinking. The way in which politicians (allegedly) manipulate statistics is a case in point. Irvine, Miles and Evans offer these observations on the subject:

> As the outright falsification of figures is risky, it is very rare. More common tactics include non-publication ('the truth, nothing but the truth, and as little of the truth as possible'), delay of publication (such as the six-month delay of 'race' statistics by Crossman, who feared the reaction to projections of the number of 'immigrants'), misleading or inadequate commentaries on published figures (for example, in order to be 'optimistic' about the economic situation), and 'massage'.
>
> Massage is the manipulation of the way figures are derived or presented in order to create desired impressions. It covers a continuum of deception ranging from the choice of the most presentable of several equally accurate modes of presentation, to subterfuges which, in practice at least, amount to falsification. We mention four types of massage:
>
> 1. *Changing the definition of terms in order to create a desired effect.* The unemployment figures reported over the years in the *Department of Employment Gazette* reflect a successive narrowing of the definition of unemployment, to keep figures down – with regard to the inclusion or exclusion of such groups as married women, school-leavers, people out of work for short periods and part-time workers.
>
> 2. *Unjustifiable extrapolation of trends.* For example, Chancellor of the Exchequer Healey claimed in an election campaign that inflation was running at eight and half per cent per year. He could claim to be right only by ignoring seasonal adjustment and extrapolating from a short untypical part of the year; the rate of inflation at the time, taken over any reasonable period, was much higher.
>
> 3. *Manipulation of adjustments.* Here advantage is taken of the practice of adjusting figures so as to reduce the impact on them of atypical circumstances at the time of collection. An example has been the selective application, under pressure from the Treasury, of adjustments to the Retail Price Index[5] , made with the purpose of reducing the impact on the Index of rapid fluctuations in the prices of eggs, fresh vegetables and other perishable foodstuffs, when and only when such adjustments served to depress the index.
>
> 4. *Manipulation of categories.* The categories into which items (foods, people, or whatever) are divided may be chosen in order to clarify or to obfuscate. A

> relatively small group of items, which is nevertheless of interest, may be included in a larger category, or in a rag-bag category marked 'other' or 'miscellaneous' for the purpose of concealment. Thus are concealed, for example, the staffs and budgets of secret agencies, and the embarrassing survival in a handful of villages of all-age schools.
>
> ***Irvine, Miles and Evans, 1979; reprinted in Hammersley, 1993:162-3, words used instead of numerals***

This shows why researchers do not take 'official statistics' at face value, but it is not *always* the case that official statistics/reports are *designed* to deceive. It is not inconceivable that the ('official) statistics may be recording something different to what the researcher thinks they are recording. The point made in the previous chapter, (about researchers understanding the use and interpretation of 'internal' documents), can also be seen to apply here. Here again, the onus is on the researcher to ensure that they understand the nature of the information, and the message it conveys (both for themselves and their reader). For example, police statistics only show the number of offences that have been reported. By definition, these figures do not show the number of offences that go unreported, nor do they show the offences that are not considered 'crimes' (for example civil offences such as trespass). Another problem with such statistics is that they only show the number of *offences*. In other words, they do not show the number of victims, or (quite reasonably) the number of perpetrators. This means that it is not possible to tell, from police statistics, how many people have been the victims of crime, or how many people have perpetrated these crimes. These are what researchers refer to as 'hidden populations' - "Official statistics…notoriously underestimate the size of hidden populations and the extent of deviant activities" (Lee, 1993:45). There is, however, one form of publication, which *does* provide information on (through providing a voice for) 'hidden populations', and that is the(ir) (interest group) newsletter.

Interest Group Publications (Newsletters)

There is an interest group (producing newsletters and leaflets) for almost every social minority, disorder/illness, political opinion, interest, occupation, belief and hobby, imaginable. While such publications are 'public' in the sense that they can be read by anyone, they are clearly intended for a specific audience (the organisations members). As with 'official' publications, one can make this distinction between 'public' and 'for the public', which may be important when it comes to interpreting certain information. Such organisations are unlikely to exercise any formal control over the quality of the information that they disseminate, in their publications, but the nature and context of the publication means that the information is unlikely to be invalid. Interest groups are, by definition, made up of individuals with a common interest; this gives them an affinity with other members. As a result their publications (newsletters and leaflets) tend to be written: *for members, by members* (or interested individuals), as a form of social support. Given the nature of the situation, people are unlikely to *(knowingly)* mislead their 'own'. (What goes around comes around.)

This suggests that the group's publication will be a (relatively) valid source of information. Although the information is only (implied to be) valid by virtue of this norm of reciprocity (which comes from the support that members give to one another), it is, however, a basis on which to have confidence in the quality of the information. Again, it comes down to looking for reasons not to believe the source (see previous chapter). In some cases (such as with hidden populations) this may make the interest group the *most* reliable source of information on a particular subject.

Not only are interest groups useful for finding out about 'hidden populations'; they can also be a source of knowledge on their own area of interest. A few years ago, I was considering some historical research, which involved the use of court records. I can not read Latin, and so needed to find out when the courts began using English as their official language. Fortunately, I live near Kentwell Hall, which stages regular recreations of medieval life (around the year 1666). These recreations involve several hundred actors/enthusiasts, playing a variety of roles, from lords and scholars to servants. While the actors/enthusiasts allow visitors to interact with them, they remain in character while doing so, even when engaged in conversation. On a visit to one of these recreations, I took the opportunity to ask a 'legal scholar': *'how long did he think it would be before all the law courts were using English as their official language?'* (I reasoned that if this person knows enough to play the role of a medieval, legal scholar, then he should know when the courts switched from using Latin to using English and I took his answer to be authoritative.) While this may not be the most scholarly way to pursue a research question, it is good enough to provide a starting point, and it shows that 'authorities' can be found in the most unlikely of places.

Section Conclusion

The 'quality controls', cited (or implicit) in the above examples, may not be what everyone *(ordinarily)* thinks of as quality controls. Although the latter two examples can perhaps, more accurately, be described as examples of *self*-regulation (as opposed to a *system* of regulation), they still have an effect on the quality of the information. Quality controls are not defined by the formality of their procedure, but by the effect that they have on the quality of the information, and the confidence they give one in that information/author/publication. Confidence, in a particular author/publication, can, therefore, be taken as evidence of a reason to have confidence in that particular author/publication. This provides us with a way of (indirectly) making these 'quality controls' visible. Unfortunately, these 'quality controls' can not be easily manipulated (see chapter two on experiments). It is, however, possible to compare confidence in these publications, with confidence in one, which lacks any form of regulation. A good example of which is the Internet. At the time of writing, the Internet borders on publishing anarchy – it is plagued by misinformation, impersonation and publications that (were it not for the want of being able to identify the author and/or distributor) would otherwise be illegal. The user has no guarantees, *whatsoever*, that the information (on any website) is accurate, or that author is who they purport to be. If one were to ask 'users' (of the Internet) how much confidence they had in the Internet (as a provider of reliable information), their response would probably be that "it *varies*". In other words, *it depends on the 'site'*. Internet 'sites' range from the reputable, hosted by reputable sources, to the disreputable, hosted by 'unknown' sources. In which case, how does one know that a site is reputable?' The answer is, of course, through (personally) knowing the source (or through a reliable recommendation). In many respects (in terms of the pursuit of 'knowledge'), this can be seen as a step backwards – instead of being able to have confidence in the publication *per se*, one is limited to those 'sites' which one knows (first-hand) are reputable. The whole point of having such quality controls is that they enable people to have access to 'knowledge' (by being able to have confidence in the source), which is *beyond* their own personal experiences. (Hence, the Internet is a step backwards.) If we were to ask people what they thought about the Internet, and what they thought about government publications (or interest group, newsletters), these 'quality controls' would suddenly become apparent.

Published/Public Information - Verbal

There is, essentially, only one method/source which fits this description, and that is *'The (Mass) Media'*. The Media (as a source) is probably one of the *least un*adulterated source of information there is. This means that evaluating it (as a source of information), and evaluating the research which has aimed to evaluate it (as a source of information), requires a more advanced critique than has been the case so far. As with any (secondary) source, the quality of the data is always going to be an issue. This means that the points made about validity and self-interest will apply here equally. In this case, the issue is complicated by the fact that the medium may serve a *number* of interests (a point which has not escaped the notice of media sociologists). Our understanding of this issue has not been helped by the way in which some media sociologists have gone about investigating the subject. Some researchers (or groups of researchers) seem to be intent on investigating, not *whether* the (news) media is 'biased', but *how* it is 'biased'. This touches on a number of issues discussed above – the nature of authority/expertise (applied to both the media and the researcher), the evaluation of evidence (again applied to both), the evaluation of 'scientific' research, and the possibility of a researcher 'seeing' what they want to see. As such, this discussion will not only draw on (implicitly if not explicitly) the debates and issues discussed above; it will effectively bring them together. (Which is why it is *particularly* appropriate to discuss 'The Media' at this point.)

The Mass Media

Introduction

The mass media is one of the most researched topics in the social sciences. Not only is it a *source of information*; it is also a *topic of investigation* in its own right. While this discussion will focus on the mass media, *as a source of information*, it is important not to exclude research into 'the media', from this discussion, because provides an indicator as to the quality of the information. Almost every piece of (sociological) research, which takes 'the media' as its subject, can be seen as an investigation into the quality of the information disseminated *by* the media. Topics such as: the socio-political context of the media, the media's use of social icons and imagery, the media's role in perpetuating social stereotypes, the use of hidden and double meanings, the symbiotic relationship between the media and culture, and the media as propaganda, can all be seen as research into the objectivity of the information disseminated by the media. A great deal of this literature is, therefore, relevant (technically), but it is only going to be possible to look at *some* of it in this discussion. This discussion will, therefore, focus on (the literature relating to) just one area of the media – the news (and even this will be a selective discussion). The news media is a particularly apt focus for a discussion such as this because, unlike the less serious incarnations of the medium, it actually portrays itself as *an information service*. It is the boast of the news media that it is simply reporting the 'facts' as they are, without any *'artistic embellishments'*. This has made it a 'target' for social commentators (sociologists and social researchers), who, no doubt, see this as a challenge. As far as media sociologists are concerned, if 'the news' (the most objective and impartial area of the media) can be shown to contain value judgements, selectivity, and political influence, then what does this say about the rest of the media? Investigations, such as this, are clearly useful to researchers who are considering using the media as a source of information. The problem is that the *investigation* itself may not be objective! As we will see, this is a very real possibility.

There are, principally, two things to bear in mind with this (media sociology) literature. The first is that the research question is being interpreted from a sociological perspective. Sociologists are, predominantly, concerned with the study of social aggregates, which means that they are likely to be looking for the effect that certain social aggregates have on the research subject (media). Any questions, raised by their research (with regard to the quality of the information disseminated by the media), will, therefore, be raised in this context. The perspective, itself, may not be an invalid way of ascertaining the validity of the *information* (disseminated by the media), but it does raise one, rather important, question: *'to what extent do the opinions of one discipline (with regard to the nature of the subject mater) apply to other disciplines?'* (We have already seen that 'validity' can mean different things, depending of the method/source used.) This means that in order to use this approach (using research as a way of determining the validity of the information disseminated by a particular source), one has to understand the perspective from which that source has been investigated.

The second thing to bear in mind is that an investigation is only likely to be published, if that investigation finds something *of note* to say about the subject. In other words, if a piece of research finds nothing of note to say about the subject, then it is unlikely to be published. Given that, by definition, one is only ever likely to read published material, this means that one is only ever likely to read research, which *has* found something of note to say about the subject. This, in itself, encourages researchers to find *something* to say about the subject, and the work of the *Glasgow University Media Group* is a good example of this. Throughout the late 1970s and 1980s there was an ongoing debate (among media sociologists) as to whether or not television news was (is) impartial/objective/free from social and/or political influences. The principal researchers, in this area, came from a specialist research unit at the University of Glasgow *(The Glasgow University Media Group)*. There is evidence to suggest (as will be discussed, later in this section) that the Glasgow University Media Group did not set out to investigate *whether* the (television) news was 'biased'; they set out to investigate *how* it was biased. In other words, they *were looking for* 'biases' in the news. Clearly, these debates (and the issues surrounding the Glasgow Group's research) have a bearing on the approach suggested here – if the research, which questions the objectivity of the source, is not, itself, objective, then this tells us little about the validity of the source subject (media news). The reader must decide, for themselves, whether the criticisms (of the investigations) are damning enough to discount the findings of this research (negating any questions which they may raise about the validity of the information), or whether the findings are robust enough to withstand these criticisms. Clearly, one needs to look critically at this body of literature/research. Even if the research has been conducted professionally, there are at least two good reasons why the analysis may not be *completely* objective: the perspective of the discipline, and the need to find something to say about the subject. Let us now look at this literature.

News Media - As Seen by Media Sociologists:
According to media sociologists, Lee and Solomon (1992), there are a number of ways in which *'the news'* can be 'biased'/'misleading':

> ***Headline Hanky-Panky*** – "A good way to begin detecting and decoding biases in news coverage is to take a close look at how accurately headlines encapsulate the story that follows. The headline is the most important aspect of a news story, serving as an index for the reader who is 'shopping' the paper. Most people, when reading a newspaper or magazine, scan the headlines but only actually read a few articles that perk their interest. Our 'take' on the news is, therefore, largely formed by fleeting glances at headlines and subheads that may or may not accurately reflect the accompanying stories." (p.35)
>
> ***Editorials Versus News Stories*** – "Another category of media bias involves editorials that contradict news stories appearing in the same newspaper. By editorials, we mean unsigned commentary…as opposed to…opinion pieces that run with bylines…. Sometimes editorials conflict with news stories appearing the same day in the same paper. On other occasions, information reported in the news pages will be conveniently ignored in an editorial published days, weeks or even months later." (p.36)
>
> ***Loaded Language*** – "This category of media distortion is broad and far-ranging. It could include a peculiar – or politically dubious – use of a word or phrase that might stand reality on its head. Or it might entail the selective application of phrases that, when compared in different contexts, display a pattern of bias." (p.36-7)
>
> ***Unattributed Assertions*** – "Unattributed assertions often paint with a broad brush and are of questionable validity." (p.41)
>
> ***Suspicious Sources*** – "Occasionally reporters need to speak with sources 'off the record' or on a 'not-for-attribution' basis. This makes it possible for people to provide news without risking their job, safety or reputation. Sources need to know they can trust a journalist before sharing confidentialities. This is especially true for whistleblowers in government or corporate spheres.
>
> Unfortunately, journalists often grant anonymity to people who should be publicly accountable. Instead of a bond of trust, the relationship becomes an

indentured contract, as reporters look to government and corporate elites for their daily dose of news." (p.42)

Burying the Lead – "Even when journalists include a range of sources in their reports, bias is sometimes conveyed by the placement of important sources or information. A lengthy TV news report with a particular slant might go on for a few minutes, only to be contradicted by a significant statement that is buried as an afterthought or a throwaway line at the tail end of the story. A newspaper might run a long story with an addendum or brief passage buried at the bottom that flies smack in the face of everything previously reported." (p.44)

Photographic Sleight of Hand – "Manipulative photo journalism can assume various guises. Photographs might have erroneous captions, or two photos might be juxtaposed in such a way as to create a misleading impression. The selection of photos is significant, for a picture sends a cue about how to view an article before we actually read it." (p.46-7)

Passive Phrases and Gloss-Over Euphemisms – "This linguistic technique can impart a subtle bias to news coverage. While the press sometimes has a penchant for hype and sensationalism, reporters and editors frequently choose words that ends up neutralizing events that are controversial. Euphemistic phraseology can covey a skewed impression about current events by blurring realities." (p.49)

The Numbers Racket – "The numbers racket is a category of media bias that includes rigged statistics, inflated or deflated estimates of attendance at political demonstrations, contradictory tallies and numerical tricks." (p.52)

Polls Made to Order – "Polling data is often deceptive. Results can be skewed in one direction or another depending on what questions are asked and what procedures are used to gather data. This may explain why polls on the same subject may yield markedly different results." (p.54)

After Lee and Solomon, 1992

While these observations seem to be reasonable enough, there is something disconcerting about the way in which Lee and Solomon portray the (news) media, and their analysis of it:

> Reading between the lines and seeing through media con games isn't a natural talent, it's an acquired skill. Obviously if a person has expertise in a particular area, he or she will be better equipped to ascertain the strengths and weaknesses of news coverage of that subject. However, most of us aren't experts in the nuances of international diplomacy, military strategy, corporate mergers and leverage buyouts, or a hundred other topics that media report on regularly. "Of course, it is possible for any citizen with time to spare, and a canny eye, to work out what is actually going on," said novelist Gore Vidal, "but for many there is no time, and the network news is the only news even though it may not be news at all but only a series of flashing fictions intended, like the avowed

commercials, to keep docile huddled masses, keep avid for products addled consumers."

Lee and Solomon, 1992:32

This quote conveys the impression that the media is in the business of deceiving people, and that the general public is not intelligent enough to see it. Expressions such as 'con games' (and from the quote), flashing fictions' and 'docile huddled masses' all reinforce the idea that the media is 'conning' the unsuspecting public. Straight away, this alerts one to the possibility that Lee and Solomon's analysis may not have been entirely objective. Perhaps more of a concern is the idea that one needs specialist training, in order to see through these deceptions ('con games'). In other words, one needs to know *how* to look at the media. There are many methodologies (and approaches) which would, no doubt, have something to say on this subject, but, as far as this discussion is concerned, the thing to note, here, is the way in which this suggestion comes dangerously close to the idea of *'looking for'* supporting evidence – which can then become a self-fulfilling prophecy (see chapter one). As a side issue, Lee and Solomon also seem to under estimate the critical capabilities of 'the audience' – it is not that difficult to spot the political leanings of most British newspapers. (Nor do they seem to have allowed for the possibility that the docile huddled masses may also be educated enough to question what kind of analysis they are likely to be getting from researchers, who begin their discussion with such a statement.) Many of the sentiments, expressed in this quote, can also be seen in the work of the Glasgow University Media Group.

In the late 1970s, the University of Glasgow's Sociology Department formed a specialist media research group – the Glasgow University Media Group. Their research (into news reporting at the BBC) yielded three (principal) publications: *'Bad News'* (1976), *'More Bad News'* (1980), and *'Really Bad News'* (1982). At the time, the group's publications, and the way in which they went about conducting their research, caused a certain amount of controversy. However, they were not the only ones conducting research, into the BBC's news service, at this time. Philip Schlesinger (*'Putting "Reality" Together: BBC News'*, 1978), and the Annan Committee (a government commission) (*'HMSO Report of the Committee on the Future of Broadcasting'*, 1977) were also conducting research into the BBC's news services. These publications provide useful comparisons, from which it is possible to assess the validity of the Glasgow Group's methods and conclusions. There are also a number of publications, written by members of the academic community, criticising the Glasgow Group's approach (Boyd-Barrett and Braham, *'Media Knowledge and Power'*, 1987; Harrison, *'TV News: Whose Bias?'*, 1985; Whitehead, *'ITN Attacks on Glasgow Front'*, 1985), as well as papers, written by members of the group, defending their research (for example, Philo, *'Whose News?'*, 1987).

'Bad News' does not just report on the research of the Glasgow Group, it actively constructs a view of television news professionals, through its portrayal of their 'reception' to the Glasgow Group's research/ers. At various points in the book (page references found in *'Bad News'*), reference is made to the broadcasters' hostility (pp.52 and 56), and defensiveness (pp.ix and 56), the journalists' shyness at being questioned (p.58), the senior executives' distrust of the group, the various publications produced in response to

the Glasgow Group's work (p.57 and Philo, 1987), and the deterioration in the relationship between the researchers and the broadcasters (p.56). Mysteriously (for the Glasgow Group), all of this flies in the face of the broadcasters' "public attitude to academic research" (p.57). The explanation offered by the Glasgow Group/Hoggart is as follows: "the strength of their (the broadcasters) defensiveness is in direct proportion to their refusal to take a good look at just what they are doing each day" (p.ix).

We might be able to accept the Glasgow Group's interpretation of events, were it not for a number of things which call into question the group's characterisation of the broadcasting profession. (This suggests that they may have simply been seeing a reaction to their own approach.) The Glasgow Group certainly seemed to be unreflexive and unselfconscious, when it comes to considering what effect they (and their approach) may have had on the people they were studying. The Annan Committee report made the comment that it is possible to detect "an initial bias" in the group's work (*HMSO Report of the Committee on the Future of Broadcasting*, 1977:272 see also Harrison, 1985:10, present author's emphasis). If this is the impression conveyed by the group's report, then it may also have been the impression conveyed (to the BBC) by the group's researchers. (Hence, the reaction they saw (provoked?) in the broadcasters.) The Glasgow Group's publication quotes a statement, made by Desmond Taylor (then editor of BBC News and Current Affairs), which appears to be hinting at the same:

> "If one of you were to go into a BBC newsroom and start talking to the journalists about their editorial responsibility, they might well scent pretension and shy away."
>
> ***Glasgow Group, 1976:58***

The group's 'bad experiences' (with the BBC) were not echoed by Philip Schlesinger, who conducted a similar study into the BBC's news service. Schlesinger was not uncritical of the BBC, but his book did not provoke such a violent reaction from broadcasters. A significant factor may have been that Schlesinger conducted most of his research before the Glasgow Group even approached the BBC. It is possible that, at the time the Glasgow Group approached the BBC, they may have become tired of researchers asking them questions, or decided to change their policy. (At the time, there may have been several (on-going) research projects, conducted under the remit of the Annan Committee.) Alternatively, the BBC may have been more benevolent to Schlesinger, because his research was for his Ph.D. (Schlesinger, 1978 acknowledgements). While it is always possible that Schlesinger did not fully report any difficulties he may have had, the extent of the co-operation afforded to him (1260 hours of observation and 120 hours of interviews, over four years, ibid:13), it does not sound like he had that many difficulties. Even though the Glasgow Group had a relatively better relationship with ITN, they could still explain other researchers having an *even better* relationship (with ITN) by recourse to ulterior motives (Philo, 1987). It is not much of an exaggeration to say that the picture, painted by the Glasgow Group, is little short of a conspiracy against them.

'Bad News' not only portrays the broadcasting profession in a bad light, it makes near unfounded statements about what are, supposedly, "common-sense assumptions", "widely

held views", and "supporting evidence". All the following assertions (taken from 'Bad News') were made without any supporting evidence, or with evidence which was of little relevance to the issue in question:

> *At some common-sense level television news has to appear neutral or its credibility would evaporate. (p.1)*
> *There is the agreement widely shared in our society that television news is more objective than the press. (p.1)*
>
> *A number of observers have noted the 'sameness' of the news from day to day. (p.84)*
>
> *In recent years there has been growing criticism from both sides of the industry as to the nature of television news. (p.40)*

It may well be that these statements represent genuinely held convictions, or impressions gained from their social contacts, or literature reviews and seminars conducted on the subject (Glasgow Group, 1976:50). Alternatively, they may be just a way of expressing the group's beliefs and opinions. However, the way in which these views are presented gives the impression that the researchers are mere by-standers to the views they report. There appears to be no overt recognition, by the group, that these views may simply be their (the group's) *own* opinions. When questioned about the group's research, Greg Philo (a member of the Glasgow Group) said: "As scientists, it was our job to simply measure what was going out on television." (reported in Harrison, 1985:21)

Ironically, the Glasgow Group make a number of comments (about the broadcasters), which come very close to the kind of comments, which philosophers and sociologists (of science) have made about scientists (and social scientists) (see for example Barnes, 1985; Mulkay, 1979: Woolgar, 1988). With very little imagination, one can envisage the following statements, *made by* the Glasgow Group, being *applied to* the Glasgow Group. (All quotes are from *'Bad News'*.)

> *To the extent that broadcasting institutions are trading on some taken for granted consensus that they are in the business of shaping public consciousness. The sources for this hidden consensus...lie in the political culture of our society. (p.12)*
>
> *With few exceptions they (the workers in the newsroom) were at pains to assure us that the news was undistorted and uncensored and was governed by journalistic criteria and expertise...the participants themselves become unaware of the many factors including the obvious cultural ones that govern and shape their output. (p.60)*
>
> *(T)he ideology of news producers and reporters is one that lays great stress on neutrality. This is expressed in a series of professional constraints and practices backed by a commitment to impartiality and balance. (p.39)*

The criticism (of the group), being made here, is not that the group is making such assertions, but that they are doing so, unselfconsciously and unreflexively. It difficult to imagine how a group of experienced researchers could make statements such as these (about members of another profession), without pausing to consider whether they might not apply to their own profession. Alternatively, the group may have seen themselves as redressing some kind of social injustice, which entitled them to more of a free hand with their evidence. (Or, perhaps, they just hoped that no one would notice.)

The group also conveys the impression that (they believe), if anyone were to examine their data in full, then they would come to exactly the same conclusions. They tend to attribute differences, in interpretation, to the reader/other researcher not being *fully* acquainted with their data. For example, in the foreword to *Bad News*, Richard Hoggart says:

> Even before publication it has had a hostile or at best suspicious reception from many people involved in that (television news) process. this was partly because the group issued a brief report of its findings to the Annan Committee on Broadcasting, a report which – because it had to be put together quickly – could not give all the evidence in support of what it said.
>
> ***Hoggart/Glasgow Group, 1976:ix***

From the comments made by the Annan Committee, it is clear that the evidence, presented in *'Bad News'*, *is* sufficient, and that the committee *could* appreciate the points, being made by the Glasgow Group (see para. 7.17 of the report). However, what the Glasgow Group failed to appreciate was that the Annan Committee was not concerned with cultural biases, of which the journalists were "unaware" (the main thrust of the Glasgow Group's criticism) (Glasgow Group, 1976:60), but the reporting of stories with "deliberate and calculated bias" (*HMSO Report of the Committee on the Future of Broadcasting*, 1977:272).

While there can be little doubt that the media sociologists have a valid argument, the real question (as far as this discussion is concerned) is, perhaps: *'how significant are these 'biases'* (from the perspective of someone wishing to use this information)?' In other words, could these media sociologists be making this more of an issue than it really is?' This is how Sigal sees the (news media's) situation:

> Readers, whether they are attentive citizens or interested public officials, tend to lose sight of the fact that news is not reality, but a sampling of sources' portrayals of reality, mediated by news organizations. To coordinate the activities of their staffs with a modicum of efficiency, newspapers can do little more than establish some standard operating procedures for sampling potential sources. Whatever procedure they adopt unavoidably biases their selection of content. While no procedure can assure a sampling to satisfy all readers, some procedures affect the likelihood that some points of view and some people will be systematically excluded from press coverage. While no procedure can exclude outright lies, stories that sample only one source or one group of like-minded sources are almost certain to be partial.
>
> ***Sigal, 1986:27-8***

What Sigal seems to be suggesting here, quite reasonably, is that the news media has to be make decisions as to what (and perhaps who) constitutes 'news' (what is important), as well as what sources to take notice of, and it probably has established guidelines to this effect. If these organisations were to make their guidelines available (to the public, or interested parties), then this would make editorial decisions more understandable, and this would help to temper the (media sociologist's) criticisms that they are 'biased'. Surely, if the news media does not agree with the way in which certain social commentator's have deconstructed their product, then all they have to do is show the public (and the social commentators) how wrong the media sociologists are, by showing the public how it is actually done! Until they are willing to do this, the media sociologists' view is the best interpretation available. This begs the question: *'why are these organisations not more open about their editorial policies?'* Perhaps they are afraid that their policies will be misconstrued as 'biases', or that unscrupulous celebrities may use them to increase their own 'exposure'. Here, the news media is overlooking one, very important, issue: *how people appraise 'knowledge'*. Scientific knowledge is valued because the process is amenable to public scrutiny, therefore, if the news media wants its 'knowledge' to have the same status, then it will have to make its information gathering process amenable to public scrutiny. The crucial factor is not the objectivity of the process; it the fact *that it can be seen* to be objective (thus giving people confidence in the knowledge).

Using the Media as a Source of Information - A Conclusion

This section began with a statement of the 'problem' to the effect that the news media portrays itself as an *objective* 'information service', when it is not, and researchers working in this area seem to be intent on showing that the news media is 'biased'. The news media's boast can be seen not only as an attempt to bolster its own profession's 'reputation', but as an attempt to lay claim to a quality of knowledge which is usually only found in academic research (see Taylor 2007 for a discussion). In other words, it is 'treading on the toes' of the academic community. This may explain why some media sociologists seem so to be so intent on defeating the news media's claim to objectivity. Both sides (the news media and media sociologists) appear not to understand the other's profession. The news media is clearly unaware of what gives scientific knowledge its status (mistakenly believing that it is a claim to objectivity), and the media sociologists seem to be confusing 'editorial decisions' and 'journalism' with 'biases'. The (news) media can not claim that its knowledge has the same (or an equivalent) status to scientific knowledge (which is effectively what it is doing), because its approach is not 'scientific'. Being 'scientific' does not mean 'being objective' (even if one is, or can be), it means making the information gathering process visible (in order for it to be appraised). It is often the case that 'being scientific' means showing that one is *not* being objective – by demonstrating an *awareness* of the how one may be effecting the research subject (and how this may effect what one is 'seeing' and/or 'reporting') (see Taylor, 2007, in particular, the section on Feminism). In doing so, one is making the knowledge gathering process amenable to public scrutiny, and also pre-empting accusations that one has not taken all the factors into account. This *is* what it means to be scientific, and it is, clearly, not the way that journalists approach the task of getting a 'story' (as yet). It is often the case that a journalists will deliberately *look for* 'an angle' (a particular 'slant') on the story.

The news media has frequently had its practices and reporting called into question, not just by academics, but by regulators (watchdogs) and aggrieved readers. As we have seen, academics have tended to focus on the medium's (alleged) 'biases'. While the media sociologists, no doubt, have a point, could the journalists' claims of 'objectivity' not be, somewhat, 'tongue in cheek', and the media sociologists have taken this claim of 'objectivity' a little too seriously? Even to an 'unenlightened' observer, it is obvious that the 'tabloid' newspapers have a different agenda to the 'broadsheet' newspapers. While this may be the case, media sociologists would, no doubt, wish to point out that their research makes visible the less obvious 'deceptions'. In doing so, these researchers are doing what sociologists do best – investigating the effects of macro-social constructs, such as politics, culture and social control, more specifically: who controls/influences the media, the media's 'role' in modern society and the portrayal of social groups. Nonetheless, in attempting to show that the news media is 'biased', these academics have ignored some of the tenets of good practice, of their own profession.

It is unlikely that the news media will ever adopt the practice of making its information gathering visible. This means finding a way of using the information in a way which does not rely on the objectivity of the information, or even makes use or this particular feature of the medium. One way of doing this would be to look at the way in which the *same story* is reported by *different (say) newspapers*. This provides a 'composite' account of the story, as well as a way of making the various perspectives visible. While it may not be possible to see the perspective taken by one newspaper, if this is all one has, when one can compare the perspectives of different newspapers (reporting the same story), it makes *each* of them visible. This is a technique which I have used to good effect in my own research paper which can be found at the back of this book.

Bibliography for Chapters 1 - 5

Media

Almond, M. *Suspect Science*. Mail on Sunday, 9th July 1995.

BBC Broadcast (Nov 2007) *Edwardians in Colour: The Wonderful World of Albert Kahn.*

Castro, T. (2006) *Archives de la Planète.* www.ejumpcut.org/archive/jc48.2006/KahnAtlas/index.html

Neuendorf, K. A. (2001) *Content Analysis Guidebook* companion website http://academic.csuohio.edu/kneuendorf/content

Plomin, J. *It's a Stick-Up*. Guardian Education, 5th June 2001

Books, Journals, and Reports

Adler, P. (1985) *Wheeling and Dealing: An Ethnography of an Upper-Level Drug Dealing and Smuggling Community*. New York: Columbia University Press.

Adler, P. A. and Adler, P. (1993) Ethical issues in self-censorship: ethnographic research on sensitive topics. In C. M. Renzetti and R. M. Lee (Eds.) *Researching Sensitive Topics*. London: Sage.

Agar, M. H. (1980) *The Professional Stranger: An Informal Introduction to Ethnography*. San Diego, California: Academic Press (Harcourt Brace Jovanovich).

Aguirre, B. E., Quarantelli, E. L. and Mendoza, J. L. (1988) The collective behavior of fads: the characteristics, effects, and careers of streaking. *American Sociological Review, 53(Aug)*: 569-584.

Ajzen, I. (1991) The theory of planned behavior. *Organizational Behavior and Human Decision Processes, 50*: 179-211.

Allen, G. E. (1986) The Eugenics Record Office at Cold Spring Harbor, 1910-1940: An essay in institutional history. *Osiris, 2nd Series, 2*: 225-264.

American Psychiatric Association (1987) *Diagnostic and Statistical Manual of Mental Disorders, Third Edition – Revised.* Washington, DC: American Psychiatric Association

Annan Committee (1977) *Report of the Committee on the Future of Broadcasting*. London: HMSO.

Asher, S. R. and Allen, V. L. (1969) Racial preference and social comparison processes. *Journal of Social Issues, 25(1)*: 157-166.

BBC (1993) *Suggers, Fruggers and Data Muggers* (text adapted from the 'Horizon' programme). London: Broadcasting Support Services.

Bartlett, F. C. (1950 edition) *Remembering: A Study in Experimental and Social Psychology*. Cambridge: Cambridge University Press.

Barnes, B. (1985) *About Science*. Oxford: Basil Blackwell

Berger, P. L., Berger, B. and Kellner, H. (1974) *The Homeless Mind: Modernization and Consciousness*. Harmondsworth, Middlesex: Penguin.

Berkowitz, L. (1986) *A Survey of Social Psychology, Third Edition*. New York: CBS Publishing.

Berlin. B. and Kay. P (1991) *Basic Color Terms: Their Universality and Evolution*. Berkeley, California: University of California Press.

Blumberg, P. (1968) *Industrial Democracy: The Sociology of Participation*. London: Constable.

Boring, E. G. (1969) Perspective: artifact and control. In R. Rosenthal and R. L. Rosnow (Eds.) *Artifact in Behavioral Research*. London: Academic Press.

Bourne, L. E. Jr, and Ekstrand, B. R. (1985) *Psychology Its Principles and Meanings*, Fifth Edition. New York: Holt Reinhart and Winston.

Bower, G. H., Monteiro, K. P. and Gilligan, S. G. (1978) Emotional mood as a context for learning and recall. *Journal of Verbal Learning and Verbal Behavior, 17*: 573-585.

Boyd-Barrett, O. and Braham, P. (1987) *Media, Knowledge and Power*. London: Croom-Helm.

Brajuha, M. and Hallowell, L. (1986) Legal intrusion and the politics of fieldwork: the impact of the Brajuha case. *Urban Life, 14(4)*: 454-478.

British Cartoon Archive (formerly Centre for the Study of Cartoons and Caricature. (c2001) *Information for Users*. Canterbury, Kent: University of Kent.

Brook. L. (1983) Postal survey procedures. In G. Hoinville and R. Jowell (Eds.) *Survey Research Practice*. Aldershot, Hants: Gower.

Bulmer, M., Bales, K. and Sklar, K. K. (1991) *The Social Survey in Historical Perspective 1880-1940*. Cambridge: Cambridge University Press.

Burgess, R. G. (1993) *In the Field: An Introduction to Field Research*. London: Routledge.

Cannon, S. (1989) Social research in stressful settings: difficulties for the sociologist studying the treatment of breast cancer. *Sociology of Health and Illness, 11(1)*: 62-77.

Carson, R. C., Butcher, J. N. and Coleman, J. C. (1988) *Abnormal Psychology and Modern Life, Eighth Edition*. Glenview, Illinois: Scott, Foresman & Co.

Chambliss, W. (1975) On the paucity of original research an organized crime: a footnote to Galliher and Cain. *American Sociologist, 10(Feb)*: 36-39

Cherry, C. (1957) *On Human Communication: a Review, a Survey, and a Criticism*. Cambridge, Massachusetts: MIT Press/New York: John Wiley.

Childers, T. L. and Ferrell, O. C. (1979) Response rates and perceived questionnaire length in mail surveys. *Journal of Marketing Research, 16(Aug)*: 429-31.

Clark, K. B. and Clark, M. P. (1947) Racial identification and preferences in Negro children. T. M. Newcomb and E. L. Hartley (co-chairs of the Committee on the Teaching of Social Psychology of the Society for the Psychological Study of Social Issues). *Readings in Social Psychology*. New York: Holt, Rinehart & Co.

Coates, J. (1987) *Women, Men and Language: A Sociolinguistic Account of Sex Differences in Language*. London: Longman.

Coleman, P. G. (1991) Ageing and life history: the meaning of reminiscence in later life. In S. Dex (Ed.) *Life and Work History Analyses: Qualitative and Quantitative Developments*. London: Routledge.

Conway, M. A. (1990) *Autobiographical Memory: An Introduction*. Milton Keynes: Open University Press.

Courtenay, G. (1983) Questionnaire construction. In G. Hoinville and R. Jowell (Eds.) *Survey Research Practice*. Aldershot, Hants: Gower.

Davenport, C. B. and Laughlin, H. H. (1915) How to make a eugenical family study. *Eugenics Record Office Bulletin, No 13.*

Davenport, C. B. (1916) Preface. In Finlayson, A. W. *Eugenics Record Office Bulletin, No 15.*

Donnelly, P. (1981) Running the gauntlet: the moral order of pornographic movie theaters. *Urban Life, 10(3)*: 239-264.

Dunn, R. (1986) Television, consumption and the commodity form. *Theory, Cultureand Society, 3(1)*: 49-64.

Dyer, G. (1992) *Advertising as Communication*. London: Routledge.

Everhart, R. B. (1977) Between stranger and friend: some consequences of 'long term' fieldwork in schools. *American Educational Research Journal, 14(1)*: 1-15.

Featherstone, M. (1982) The body in consumer culture. *Theory, Culture and Society,1(2)*: 18-33.

Finlayson, A. W. (1916) The Dack family: a study in hereditary lack of emotional control. *Eugenics Record Office Bulletin, No 15.*

Foster, J. (1990) *Villains: Crime and Community in the Inner City*. London: Routledge.

Foucault, M. (translated by Richard Howard) (1965) *Madness and Civilisation: A History of Insanity in the Age of Reason*. London: Tavistock.

Glasgow University Media Group (1976) *Bad News*. London: Routledge and Kegan Paul.

Glasgow University Media Group (1980) *More Bad News*. London: Routledge and Kegan Paul.

Godfrey, R. (Ed.) (1998) *Letters From a Prince: Edward, Prince of Wales to Mrs Freda Dudley Ward March 1918-January 1921*. London: Little, Brown and Company.

Goffman, E. (1986) *Stigma: Notes on the Management of Spoiled Identity*. Harmondsworth, Middlesex: Penguin Books Ltd.

Graziano, A. M. and Raulin, M. L. (1989) *Research Methods: A Process of Inquiry*. New York: Harper and Row.

Gregor, A. J. and McPherson, D. A. (1966) Racial attitudes among white and Negro children in a deep-south standard metropolitan area. *Journal of Social Psychology, 68*: 95-106.

Gregory, R. L. (1986) *Eye and Brain: The Psychology of Seeing, Third Edition*. London: Weidenfeld and Nicholson.

Hakim, C. (1993) *Research Design: Strategies and Choices in the Design of Social Research*. London: Routledge.

Hammersley, M. (1989) *The Dilemma of Qualitative Method: Herbert Blumer and theChicago Tradition*. London: Routledge.

Hammersley, M. (1992*) What's Wrong With Ethnography: Methodological Explorations*. London: Routledge.

Hammersley, M. and Atkinson, P. (1992) *Ethnography: Principles in Practice*. London: Routledge.

Hansen, E. C. (1977) *Rural Catalonia Under the Franco Regime: The Fate of Regional Culture Since the Spanish Civil War*. Cambridge: Cambridge University Press.

Hansen, M. (1991) Pleasure, ambivalence, identification: Valentino and female spectatorship. In C. Gledhill (Ed.) *Stardom: Industry of Desire*. London: Routledge.

Harman, L. D. (1985) Acceptable deviance as social control: the cases of fashion and slang. *Deviant Behavior, 6*: 1-15.

Harris, T. (1991) The building of popular images: Grace Kelly and Marilyn Monroe. In C. Gledhill (Ed.) *Stardom: Industry of Desire*. London: Routledge.

Harrison, B. and Lyon, S. (1993) A note on ethical issues in the use of autobiography in social research. *Sociology, 27(1)*: 101-109.

Harrison, M. (1985) *TV News: Whose Bias?* Hermitage, Berks: Polity Press.

Hippler, H-J. and Schwarz, N. (1986) Not forbidding isn't allowing: the cognitive basis of the forbid-allow asymmetry. *Public Opinion Quarterly, 50*: 87-96.

Hraba, J. and Grant, G. (1970) Black is beautiful: A reexmination of racial preference and identification. *Journal of Personality and Social Psychology, 16(3)*: 398-402.

Irvine, J., Miles, I. and Evans, J. (Eds.) (1979) *Demystifying Social Statistics*, London: Pluto Press. Reprinted in M. Hammersley (Ed.) (1993) *Social Research: Philosophy, Politics and Practice*, as 'How official statistics are produced: views from the inside'. Milton Keynes: Open University Press/ London: Sage.

King, R. (1978) *All Things Bright and Beautiful? A Sociological Study of Infants' Classrooms*. Chichester, John Wiley and Sons.

Langlois, Ch V. and Seignobos, Ch. (translated by G. G. Berry) (1898) *Introduction to the Study of History*. London: Duckworth.

Laurence, J-R. and Perry, C. (1983) Hypnotically created memory among highly hypnotizable subjects. *Science, 222*: 523-4.

Leahey, T. H. (1987) *A History of Psychology: Main Currents in Psychological Thought, Second Edition*. Englewood Cliffs, New Jersey: Prentice-Hall.

Lee, M. A. and Solomon, N. (1992) *Unreliable Sources: A Guide to Detecting Bias in News Media*. New York: Lyle Stuart.

Lee, R. (1993) *Doing Research on Sensitive Topics*. London: Sage.

Leiss, W., Kline, S. and Jhally, S. (1990) *Social Communication in Advertising: Persons, Products and Images of Well-Being, Second Edition*. London: Routledge.

Levy, E. (1988) *John Wayne: Prophet of the American Way of Life*. Metuchen, New Jersey: The Scarecrow Press..

Lewis, A. (1992) Group child interviews as a research tool. *British Educational Research Journal, 18(4)*: 413-421.

Liebow, E. (1967) *Tally's Corner: A Study of Negro Streetcorner Men*. Boston, Massachusetts: Little, Brown & Co.

Loftus, E. F. (1993) The reality of repressed memories. *American Psychologist, 48(5)*: 518-537.

Maas, H. S. and Kuypers, J. A. (1974) *From Thirty to Seventy*. San Francisco: Jossey- Bass.

Mayhew, P., Maung, N. A. and Mirrlees-Black, C. (1993) *The 1992 British Crime Survey: Home Office Research Study No 132*. London: HMSO.

McCann, G. (1988) *Marilyn Monroe*. Cambridge: Polity Press.

Media Culture and Society (Journal) *Notes for Authors*.

Miller, N. and Morgan, D. (1993) Called to account: the CV as an autobiographical practice. *Sociology, 27(1)*: 133-143.

Mitchell, J. C. (1983) Case and situation analysis. *Sociological Review, 31(1)*: 187- 211.

Morris, D. (1980) *Manwatching*. St Albans, Herts: Triad/Panther.

Mulkay, M. (1979) *Science and the Sociology of Knowledge*. London: George Allen and Unwin.

Nederhof, A. J. (1983) The effects of material incentives in mail surveys: two studies. *Public Opinion Quarterly, 47*: 103-111.

Neuendorf, K. A. (2001) *The Content Analysis Guidebook*. Thousand Oaks, California: Sage.

Neugebauer, R. (1979) Medieval and early modern theories of mental illness. *Archives of General Psychiatry, 36(Apr)*: 477-483.

Neugebauer, R. (1989) A doctor's dilemma: the case of William Harvey's mentally retarded nephew. *Psychological Medicine, 19*: 569-572.

Orne, M. T. (1969) Demand characteristics and the concept of quasi-controls. In R. Rosenthal and R. L. Rosnow (Eds.) *Artifact in Behavioral Research*. London: Academic Press.

(The) People of Thurcroft (1986) *Thurcroft, A Village and the Miners' Strike: An Oral History*. Nottingham: Spokesman.

Philo, G. (1987) Whose news? *Media, Culture and Society, 9*: 397-406.

Philo, G., Hewitt, J., Beharrell, P. and Davis, H. (1982) *Really Bad News*. London: Writers and Readers.

Pitts, M. (1987) Personality. In J. Radford and E. Govier (Eds.) *A Textbook of Psychology*. London: Sheldon Press.

Platt, J. (1981a) Evidence and proof in documentary research 1: some specific problems of documentary research. *Sociological Review, 29(1)*: 31-52.

Platt, J. (1981b) Evidence and proof in documentary research 2: some shared problems of documentary research. *Sociological Review, 29(1)*: 53-66.

Plummer, K. (1983) *Documents of Life: An Introduction to the Problems and Literature of a Humanistic Method*. London: George Allen and Unwin.

Polsky, N. (1967) *Hustlers, Beats, and Others*. Chicago: Aldine.

Popper, K. R. (1976) *Conjectures and Refutations: The Growth of Scientific Knowledge*. London: Routledge and Kegan Paul.

Posener, J. (1982) *Spray It Loud*. London: Routledge.

Prosser, J. (1992) Personal reflections on the use of photography in an ethnographic case study. British *Educational Research Journal, 18(4)*: 397-411.

Ronai, C. R. and Ellis, C. (1989) Turn-ons for money: interactional strategies of the table dancer. *Journal of Contemporary Ethnography, 18(3)*: 271-298

Rosenthal, R. and Rosnow, R. L. (1969) The volunteer subject. In R. Rosenthal and R. L. Rosnow (Eds.) *Artifact in Behavioral Research*. London: Academic Press.

Schlesinger, P. (1978) *Putting 'Reality' Together: BBC News*. London: Constable.

Schuman, H. and Presser, S. (1981) *Questions and Answers in Attitude Surveys: Experiments on Question Form, Wording, and Context*. New York: Academic Press.

Scott, J. (1990) *A Matter of Record: Documentary Sources in Social Research*. London: Polity Press.

Sigal, L. V. (1986) Sources make the news. In R. V. Manoff and M. Schudson (Eds.) *Reading the News*. New York: Pantheon.

Smyth, M. M., Morris, P. E., Levy, P. and Ellis, A. W. (1987) *Cognition in Action*. London: Lawrence Erlbaum Associates Ltd.

Spindler, G. D. (1974) Preview. In G. D. Spindler (Ed.) *Education and Cultural Process: Toward an Anthropology of Education*. New York: Holt, Rinehart and Winston.

Spradley, J. P. (1979) *The Ethnographic Interview*. Fort Worth: Holt, Rinehart and Winston.

Steinhoff, J., Pechel, P. and Showalter, D. (1991) *Voices From the Third Reich: An Oral History*. London: Grafton (Harper Collins).

Stewart, D. W. and Shanidasani, P. N. (1990) *Focus Groups: Theory and Practice*. London: Sage.

Sudman, S. and Bradburn, N. M. (1989) *Asking Questions*. San Francisco: Jossey- Bass.

Tabachnick, B. G. and Fidell, L. S. (1989) *Using Multivariate Statistics, Second Edition*. New York: Harper Collins.

Taylor, M. S. (2007) *Interacting with Statistics: The Role of Statistics in the Research Process*. Gt Shelford, Cambridge: Cambridge Academic.

Thomas, W. I. and Znaniecki, F. (1958 edition) *The Polish Peasant in Europe and America*. New York: Dover.

Vidich, A. J. and Bensman, J. (1968) *Small Town in Mass Society: Class, Power and Religion in a Rural Community*. Princeton, New Jersey: Princeton University Press.

Walker, A. (1966) *The Celluloid Sacrifice: Aspects of Sex in the Movies*. London: Michael Joseph.

Whitehead, P. (1985) ITN attacks on the Glasgow front. *New Statesman, 15th November 1985*.

Whitten, N. E. Jr. (1970) Network analysis and processes of adaptation among Ecuadorian and Nova Scotian Negroes. In M. Freilich (Ed.) *Marginal Natives: Anthropologists at Work*. New York: Harper and Row.

Whyte, W. F. (1993 edition) *Street Corner Society: The Social Structure of an Italian Slum (Fourth Edition)*. Chicago: University of Chicago Press.

Williamson, J. (1993) *Decoding Advertisements: Ideology and Meaning in Advertising*. London: Marion Boyars.

Wilson, E. (1987) *Adorned in Dreams: Fashion and Modernity*. London: Virago Press.

Woolgar, S. (1988) *Science: The Very Idea*. Chichester: Ellis Horwood/London: Tavistock.

Yin, R. K. (1989) *Case Study Research: Design and Methods*. Newbury Park, California: Sage.

Zimmerman, D. H. and Wieder, D. L. (1982) The diary-diary-interview method. In R. B. Smith and P. K. Manning (Eds.) *Qualitative Methods, Volume Two of Handbook of Social Science Methods*. Cambridge, Massachusetts: Ballinger (Harper and Row).

Expanding the Method: A Revised Approach to the Analysis of Legal Cases Reported in the Press

by Mark S. Taylor

Abstract

This paper presents a revised approach to the analysis of legal cases reported in the press, originally forwarded in my M.Phil. dissertation Taylor (1997). In addition to further 'testing' the method, the paper looks at the advantages and disadvantages of using a database (administered by the *Gale Group*), and the ways in which the method may need to be adapted for large volumes of material. As before, this is demonstrated with the aid of a practical example (R v Archer, English national press, 31 May 2001 – 20 July 2001). The material is discussed with reference to the relevant legal statutes and case decisions (in the areas of perjury, perverting the course of justice, and using a false instrument), and the educational applications of the approach. Consideration is also given to the methodological issues raised by this (revised) approach.

Introduction

In my earlier paper (Taylor, 1997), I put forward an approach to analysing the press' reporting of legal cases which demonstrated how quasi-valid information could be derived from, what media sociologists have traditionally regarded as, a rather dubious source of information – the press (see, for example, Glasgow University Media Group, 1976, 1980; Lee and Solomon, 1992; Philo, Hewitt, Beharrell, and Davis, 1982). The method used in the above paper was derived from, and applied to, the press' reporting of an actual legal case: Rantzen v *Mirror* Group Newspapers (1986) Ltd and Others (see [1993] 4 All ER 975; and English national press, from 29 November 1991 to 18 December 1991, and 1 April 1993). As well as showing how a quasi-valid account of the case could be derived from press reports, the paper also showed how the relevant socio-legal issues were evident in the reporting. While the paper does much to establish the methodological credentials of the approach, it does not really address any of the *practical* problems, associated with conducting such an investigation. This is because the investigation chose a case for which the material was readily available. The holdings of academic libraries can be anything up to six months out of date, which means that if the case is less than six months old, an academic library may not have all the titles covering that period. If the case is relatively recent (less than six months old), then it may mean having to use a number of sources, *including*: public libraries, dedicated databases, web-sites, and (of course) academic libraries. These sources differ quite markedly, in terms of their reliability, and so it is important to look at the effect that this may have on the utility of the method.

The case chosen for this investigation is the 2001 perjury trial of Lord Archer (see English national press, 31 May 2001 – 20 July 2001). Not only is the case a relatively recent one, but the (press) coverage it received makes it an ideal subject for an investigation such as this. From a practical point of view, the main differences between this, and the case discussed in the previous paper (Rantzen v *Mirror* Group Newspapers (1986) Ltd and Others), is the sheer volume of material. Whereas the case study in the first paper consisted of 52 articles, of which only 37 were actually about the case (trial), the case study, in the present paper, has over 150 articles relating to the trial alone. Consequently, the case summary and topic chronology have been used in a slightly different way, in this paper. In the first paper, the topic chronology was used to provide an informative summary of the main themes of the case: *'The Scandal', 'The Case - First Day', 'The Case – Witnesses', 'The Win', 'Comments – Journalistic', 'Comments – Letters', 'Profiles', 'Appeal Case', 'Appeal Case – Legal Commentaries'*. (Those relating to the case itself being: *'First Day', 'Witnesses',* and the *'Win' /'Verdict'*.) In this paper, they have been used to set out the 'facts' of the case, as a precursor to looking at the legal and moral issues. This has meant producing a more detailed 'topic chronology', as well as re-evaluating how this information can best be presented.

From a legal point of view, there are also a number of 'technical' differences between the current case and the one discussed in the earlier paper – the current case is criminal whereas the earlier case is civil, the legal principles are more straightforward in the current case, and the non-legal issues are 'moral' rather than 'social'. In themselves, these differences do not constitute that much of a 'test' of the method. (For a discussion of how rigorous a 'test' of 'scientific' knowledge should be, see Karl Popper's *Conjectures and Refutations: The Growth of Scientific Knowledge*, 1976.) However, the current paper is not intended to be a 'test' of whether the method can be applied to a different kind of legal case; it is intended to 'test' the utility of the procedure with a more recent case. In particular, the way in which the strengths and weaknesses of different (library/database) sources can effect the utility of the method. It therefore seems only logical to begin the paper by looking at the data gathering stage of the investigation.

Gathering The Data

The approach outlined in the first paper is based on the assumption that a comprehensive review of the relevant (press) reports will produce a pseudo-valid account of events. Only during the course of this investigation did it emerged that 'comprehensive' may be a relative term. In order to see why, it is first necessary to break 'the press' down into categories. To this end, the press can be categorised into:

National daily newspapers, printed in England:
Daily Express, Daily Mail, Daily Mirror, Daily Sport, Daily Star, Daily Telegraph, Financial Times, Guardian, Independent, Sun, and Times.
(The *Financial Times* tends to specialise in business news, but it does have some general news stories.)

National daily newspapers, with Sunday editions, printed in England:
Mail on Sunday, Sunday Express, Sunday Mirror, Sunday Telegraph, and *Sunday Times.*

National Sunday newspapers, printed in England:
News of the World, Observer, and *Sunday People.*

This list of titles represents a reasonable mix of tabloids and broadsheets:

Tabloids:
Daily Mail, Daily Mirror, Daily Sport, Daily Star, Evening Standard, Express, Sun, News of the World, Sunday People, Mail on Sunday, Sunday Express, and *Sunday Mirror.*

Broadsheets:
Daily Telegraph, Financial Times, Guardian, Independent, Times, Observer, Sunday Telegraph, and *Sunday Times.*
(Although it must be said that some *Sun*day broadsheets are very tabloid.)

There are also a number of *provincial*, daily newspapers (mostly evening titles) that report national events:

Provincial daily newspapers, printed in the United Kingdom:
Birmingham Evening Mail, Birmingham Post, Coventry Evening Telegraph, (London) Evening Standard, and (Glasgow) *Daily Record.*

This means that *comprehensive* (as applied to the press' coverage of a particular legal case) may mean different things, depending on which *part* of the country (England) you come from, or even which country from (within the United Kingdom) you come from. What then constitutes *'comprehensive'* for an approach such as this? Methodologically speaking, the answer to this question must surely be the same as it is for any question about sample size – the larger the sample, the greater the probability that it will include all the required perspectives and opinions. While this is may the solution, from a methodological point of view, it is of little use when it comes to determining whether or not the loss of a title is likely to effect the validity of the investigation. This was one of the problems encountered in the course of the current investigation. Towards the end of June (2001) (four weeks into the case), I discovered that my university's (library) holdings of the *Daily Star* were several months out of date, and likely to remain that way for the foreseeable future. (This was caused by administrative problems at the agency responsible for supplying the library with back issues of the title.) At this point it appeared only prudent to begin buying the *(Daily) Star*, on a daily basis, in anticipation that the holdings for May and June (see summary and topic chronology below) would *soon* be brought up to date, and that it would still take some time for the July editions to come through.

However, the library's holdings had still not been updated by the end of the trial.
This meant that I had three options: 1) wait for the library's holdings to be updated, 2) pay a specialist search agency to find copies of the missing material, or 3) sacrifice the contribution of this title. There was no telling when the library's holdings would be updated, and the various search agencies were quoting figures of £50-£90 for the job. This meant looking seriously at the contribution of this title, with a view to eliminating it from the investigation. The topic chronology of this case (see below) shows that the reports tend to include the same kind of information (and this irrespective of whether they are tabloids or broadsheets).

Looking through those editions of the *Daily Star* that were purchased during the first two weeks of July (2001), its reports do not appear to contain anything that is not in any of the other tabloids. This means that the title does not actually contribute anything to the account of events (over and above that which can be found in the combined reports of the other titles). I therefore decided not to pursue the missing editions. For this reason, the topic chronology only shows entries for the *Daily Star*, from 29 June 2001 to 20 July 2001. While this approach may not be able to provide a definitive (practical) answer to the above question; it does show that there comes a point when the contribution of an certain title (over and above those that have already been surveyed) may no longer warrant its inclusion in the investigation.

This episode highlights one of the main difficulties when conducting an investigation such as this, namely that not all libraries hold all (of the above) titles. Local, public libraries tend to hold only a selection of titles (the most popular that their budgets will allow), and then this is only for a limited period (which can be anything from a day to several years). The decision, as to which titles to hold, and for how long, is made at the individual library level. This means that holdings may vary, even between libraries in the same geographical area. Academic libraries are not adverse to 'holding' (keeping back copies, supplied via a publisher or an agency) tabloid titles, but very few, if any, 'take' (order) them on a daily basis. This often means having to wait for the publisher to furnish them with back copies of the same, which (as discussed above) can make their holdings of certain titles several months 'behind'. Very few libraries, of any kind, 'take' or 'hold' the *Daily Sport*. This is perhaps because the *Daily Sport* is best known for its features of well endowed, semi-naked women than its news coverage. The 'availability' of certain titles is further compounded by the fact that it is not uncommon for tabloid newspapers (in particular) to 'go missing' from public libraries.

The alternative is, of course, to use a computerised database. While many of the 'quality' newspapers (broadsheets) have their own web-sites on which they reproduce (back) copies of their own articles (see, for example, the *Guardian* web-site), there is at least one organisation that maintains a general database of newspaper articles – the *Gale Group*. The information held about each article, on the *Gale Group* database, includes: the title of the newspaper, the date, the page number, the headline of the article, the reporter, the text in full, and the captions of the photographs and/or illustrations (if applicable). However, it does not contain any information about the layout of the text and heading(s), nor does it contain copies of the photographs/illustrations themselves. While this highlights one of the advantages of using the printed edition, it is only really an advantage if the report is being used as the subject of the research, rather than as a source of information. (A distinction made in the first paper, Taylor, 1997.) The *Gale Group* database is, however, by no means complete; during this case, they did not have permission to reproduce articles published in the *Guardian*, nor did they have permission to reproduce articles published in the *Daily Star* (personal communications from the *Gale Group*, 9 July 2001). There are also a number of other titles absent from their database, most notably the *Daily Express*.

The success of any database search depends on being able to identify 'key words' that uniquely identified the relevant material. In this case, a 'key-word' search of 'Archer AND Lord' and 'Archer AND Jeffrey', from 31 May 2001 (the first day of the reporting) onwards, 'found' (presumably) all of the relevant material, and little else besides. This *differentiated* 'key-word' search produced one rather unexpected, but quite interesting, finding. Namely, some of articles referred to the defendant, almost exclusively, as '*Lord* Archer', while others referred to him, almost exclusively, as '*Jeffrey* Archer'. The results of one of these *differential* 'key word' searches is shown in the table below. This was a search of national, daily newspapers (in the *Gale Group* database), from 31 May 2001 to 29 June 2001. (A 'Yes' indicates that an article was found with that key word.) As this table shows, the use of 'informal' and 'formal' titles does not break down into a 'nice' 'tabloid' (T) / 'broadsheet' (B) split. While the majority of articles refer to the defendant as *Lord* Archer, the *Independent* and *Daily Telegraph* (both broadsheets) both *tend* to refer to him as *Jeffrey* Archer. (Table continued overleaf).

Results of Differential 'Key Word' Searches from the Gale Group Database

Newspaper & Date		*'Lord'*	*'Jeffrey'*	Newspaper & Date		*'Lord'*	*'Jeffrey'*
Mail (T)	31 May	Yes		Mail (T)	13 June	Yes	
Mirror (T)	31 May	Yes	Yes	Mirror (T)	13 June	Yes	
Sun (T)	31 May	Yes		Sun (T)	13 June	Yes	
Independent (B)	31 May		Yes	Independent (B)	13 June		Yes
Telegraph (B)	31 May	Yes	Yes	Telegraph (B)	13 June	Yes	
Times (B)	31 May	Yes	Yes	Times (B)	13 June	Yes	Yes
Mail (T)	1 June	Yes		Mail (T)	14 June		
Mirror (T)	1 June		Yes	Mirror (T)	14 June	Yes	
Sun (T)	1 June	Yes		Sun (T)	14 June		
Independent (B)	1 June	Yes	Yes	Independent (B)	14 June	Yes	Yes
Telegraph (B)	1 June		Yes	Telegraph (B)	14 June	Yes	
Times (B)	1 June	Yes		Times (B)	14 June	Yes	
Mail (T)	2 June			Mail (T)	15 June	Yes	
Mirror (T)	2 June		Yes	Mirror (T)	15 June	Yes	
Sun (T)	2 June	Yes		Sun (T)	15 June	Yes	
Independent (B)	2 June		Yes	Independent (B)	15 June		Yes
Telegraph (B)	2 June		Yes	Telegraph (B)	15 June		Yes
Times (B)	2 June	Yes		Times (B)	15 June	Yes	Yes
Mail (T)	5 June	Yes		Mail (T)	16 June		
Mirror (T)	5 June	Yes	Yes	Mirror (T)	16 June	Yes	
Sun (T)	5 June	Yes		Sun (T)	16 June		
Independent (B)	5 June		Yes	Independent (B)	16 June	Yes	Yes
Telegraph (B)	5 June		Yes	Telegraph (B)	16 June	Yes	Yes
Times (B)	5 June	Yes		Times (B)	16 June	Yes	Yes
Mail (T)	6 June	Yes		Mail (T)	19 June		Yes
Mirror (T)	6 June	Yes		Mirror (T)	19 June	Yes	
Sun (T)	6 June	Yes		Sun (T)	19 June	Yes	
Independent (B)	6 June		Yes	Independent (B)	19 June	Yes	
Telegraph (B)	6 June		Yes	Telegraph (B)	19 June	Yes	
Times (B)	6 June	Yes		Times (B)	19 June	Yes	Yes
Mail (T)	7 June	Yes		Mail (T)	20 June	Yes	
Mirror (T)	7 June	Yes		Mirror (T)	20 June		Yes
Sun (T)	7 June	Yes		Sun (T)	20 June	Yes	Yes
Independent (B)	7 June		Yes	Independent (B)	20 June		Yes
Telegraph (B)	7 June		Yes	Telegraph (B)	20 June	Yes	
Times (B)	7 June	Yes	Yes	Times (B)	20 June	Yes	Yes
Mail (T)	8 June	Yes		Mail (T)	21 June		Yes
Mirror (T)	8 June	Yes		Mirror (T)	21 June	Yes	
Sun (T)	8 June	Yes		Sun (T)	21 June	Yes	
Independent (B)	8 June		Yes	Independent (B)	21 June	Yes	
Telegraph (B)	8 June		Yes	Telegraph (B)	21 June		Yes
Times (B)	8 June	Yes		Times (B)	21 June		

	12 June	Yes	
Sun (T)	12 June	Yes	
Independent (B)	12 June	Yes	Yes
Telegraph (B)	12 June		Yes
Times (B)	12 June	Yes	Yes
Times (B)	28 June	Yes	
Mail (T)	29 June	Yes	
Mirror (T)	29 June	Yes	
Independent (B)	29 June		Yes
Telegraph (B)	29 June	Yes	

There is, however, one note of caution to be sounded here, and that is: a 'hit' in both columns does not necessarily mean that the article contains both 'key' words, it may mean that there are two articles, each written by a different journalist (see below), each using a different forms of address. This also explains why a particular 'key' word may yield a 'hit', for a title, on one day, but not another. It is, therefore, important to check the headlines of the articles, and not to assume that a 'hit' is a duplicate of one found with another key word.

In a comparison of the two types of searches (database and manual), the database search is clearly better at 'finding' 'feature articles'. Feature articles are 'columns', or 'comment' pages, which give journalists, notarieties, or members of the public an opportunity to 'have their say' about a particular issue. It is not difficult to see why the computer is better at detecting these kinds of articles, when one looks at the different search strategies used by humans and computers. Feature articles are often located towards the back of an edition (or even in a separate section), away from the main news pages. In a manual search, these pages are likely to receive less attention than the 'serious' news pages. Similarly, a human (re)searcher will be on the look-out for cues (to the 'story'), such as headlines or pictures. Many feature articles do not have headlines or pictures. However, with a database search, a computer simply 'looking for' (carrying out instructions to 'match') a certain key word (or combination of key words), occurring *anywhere* in its database. This particular 'advantage' (of using a database) can, therefore, be seen to be a product of the differential search strategies used by humans and computers.

Clearly, this raises the question: *'What role do feature articles play in an investigation such as this?'* Lee and Solomon argue that 'editorials' contribute to the overall impression (of the 'story') created by the newspaper (1992). It may be argued that 'feature articles' should (perhaps) also be seen as part of the newspaper's coverage of a particular 'story'. 'Feature articles' played a more significant part in the (reporting of) the Rantzen case than they did in the (reporting of) the Archer (perjury) case (except, of course, when the case was over – see English national press, 20 July 2001 onwards). In the Rantzen case, the feature articles consisted (mainly) of legal commentaries, and 'profiles' of Ms Rantzen's career (Taylor, 1997). This may well have been a reflection of the subject (Ms Rantzen's career), and the ambiguous nature of tort law. The 'feature' articles served a more 'peripheral' function, in the Archer (perjury) case – Sam Leith's comments about Lord Archer's 'pulling power' (*Daily Telegraph*, 22 June 2001), Valerie Grove's comments about Lord

Archer's former mistress, (*Times*, 20 June 2001), Sue Caroll's comment's about Lord and Lady Archer's relationship (*Mirror*, 20 June 2001), and Naomi Marks (*Independent Sunday*, 10 June 2001) and Tim Luckhurst's (*Daily Mail*, 9 June 2001) comments about the changing role of 'personal assistants', were all very interesting, but did not contribute much to our understanding of the case. Only after the verdict was known did the press publish feature articles about Lord Archer (*Daily Express*, *Daily Mail*, *Daily Star*, *Daily Telegraph*, *Guardian*, *Mirror*, *Sun*, and *Times*, 20 July 2001 onwards). The decision, not to publish any feature articles during the course of the trial, may have been in response to a request, from the courts, not to do so. Alternatively, the press may have adopted a policy of self-censorship, in order not to prejudice the outcome of the trial. (Thus depriving Lord Archer of the opportunity to appeal on the grounds that the jurors may have been influenced by the press' coverage of the case.) Either way, this suggests that 'editorials' (feature articles) may not 'colour' the reporting of legal cases, as much as they do other news items (at least for the duration of the trial).

The (*Gale Group*) database search did, however, produce one unanticipated complication, namely that the same article may be given different headlines in different editions of the *same* newspaper. For example, an article which appears in the *Mirror*, on 31 May 2001, with the headline "Give Me An Alibi And I'll Pay You £20,000", in the database edition, has the headline "Thanks For The Alibi, I'll Give You £20,000", in the printed edition. (The text is identical in each case.) Similarly, an article which appears in the *Mirror*, on 4 July 2001, with the headline "Pal 'Duped Archer After £20,000 Insult'", in the database edition, has the headline "'Archer Insulted His Mate'", in the printed edition. The article also appears on different pages of the two editions (page 12 in the database edition, and page 16 in the printed edition). This 'problem' is not just confined to tabloids. An article which appears in the *Times*, on 19 June 2001, with the headline "Archer 'Needed String of Alibis to Cover Affair", in the database edition, has the headline "Archer Trial Told of 'Numerous Infidelities'", in the printed edition. Similarly, an article which appears the next day, also in the Times, has the headline " Archer Dumped Me to Win Top Tory Job, Says Mistress", in the database edition, but the headline "Archer's Mistress Tells Old Bailey of the End of the Affair", in the printed edition. Clearly, this is a problem if the database is being used to check the completeness of the manual search. If the (*Gale Group*) database is being used in this way, it is best to obtain a print-out of all 'citations' (summaries of the articles), and then print out the full text of any unfamiliar (from their headlines) articles.

There are, however, a number of 'methodological' advantages of using a database. When viewed from a *Positivist* perspective, this (database search), effectively, designs out one of the method's confounding variables – the search strategy of a human (re)searcher. *Positivism* sees the researcher as nothing more than an instrument of the research process (specifically a data gathering instrument), therefore, replacing a human 'searcher' with a computerised one, would be seen (by Positivists) as simply replacing one research instrument with another. The basic process remains unchanged. If anything, it is a step closer to achieving the *(Positivist)* ideal of objective, value-free knowledge. Unfortunately, the process still relies on a (human) researcher, knowing how make the best use of this (other) research tool (i.e. the database). When it comes to the issue of verification, the database may cause as many problems as it solves. While the (*Gale Group*) database gives one access to articles that one would not have otherwise known about (because they can only be found in provincial titles), it may be difficult for a person to verify the findings of an investigation that uses this material, if they do not have access to the database. For this reason, the current investigation uses only material from English, national, daily (and some Sunday) newspapers, excluding the *Financial Times*, *Daily Sport*, and the *Daily Star* (before 29 June 2001).

Legal and Moral Issues

The legal principles of the case, used in this investigation (R v Archer, English national press, 31 May 2001 – 20 July 2001), are considerably more straightforward than those of the case used in the first investigation *(Rantzen v Mirror Group Newspapers (1986) Ltd and Others [1993] 4 All ER 975)*; English national press, 29 November 1991 – 18 December 1991, and 1 April 1993; Taylor, 1997). In order to determine which (legal) decisions and issues may have a bearing on this case, it is first necessary to look at the charges against Archer and his co-defendant (Ted Francis). These charges, principally, relate to the alibi that Lord Archer had for the evening (that the *Daily Star* alleges) he spent with (the prostitute) Monica Coghlan. This alibi enabled him to successfully sue the *Daily Star* for defamation (Raphael, 1989). The charges are as follows:

Archer:

Perverting the Course of Justice (Count One):
[B]etween December 22, 1986, and April 7, 1987...[Lord Archer] procured Edward John Francis to provide his [Archer's] solicitors with a version of events in relation to the evening of September 9, 1986 which he [Lord Archer] knew to be false....

Perverting the Course of Justice (Count Two):
Between March 26, 1987, and July 7, 1987,....[Lord Archer] failed reveal the existence of his main office diary....provided Angela Peppiatt with a blank 1986 diary which he instructed her with entries to write in the diary.....[and]....handed to his solicitors for use in the court case as his main office diary.

Perverting the Course of Justice (Count Three):
Between April 6, 1987, and July 7, 1987, made entries in [a]...diary for September 1986 which purported to have been made on or before that date....

Using a False Instrument (One Count):
Between March 26, 1987, and July 7, 1987, he [Lord Archer] used the A4 appointments diary with the intention of inducing the newspaper...to accept it as genuine.

Perjury: (Count One):
On June 23, 1987, he [Lord Archer] made a statement under oath for use in judicial proceedings that the [diaries provided in court] were the only documents of that type which had been in his possession.

Perjury: (Count Two):
On 8 July 1987, as a sworn witness in judicial proceedings…knowingly made a false statement that the diary known as the main diary…was in existence and contained the entries relating to September 8 and 9 1986, before October 26, 1986.

Francis:
Perverting the Course of Justice (One Count):
[Francis] agreed to provide and later provided solicitors on behalf of Archer, a version of events….which he knew to be false and materially connected with an action against Express Newspapers….[in the libel trial].

(Daily Telegraph, 31 May 2001; Times, 31 May 2001; Daily Mail, 1 June 2001; see also the Sunday Telegraph, 13 May 2001 for a slightly less precise account of the charges.)

Let us now look at how the law defines these offences:

Perverting the Course of Justice

Attempting to pervert the course of justice is a common law offence (Stones, 1999: 8-23567). As such, the law is laid down in cases rather than statutes. While the law in this area seems somewhat inconsistent, in that "[a] course of justice must have been embarked upon….[but it]… is not necessary for a police investigation to have commenced when the acts in question occurred….It is sufficient for a conviction if the defendant intended to mislead a judicial tribunal in any or all of the possible judicial proceedings which might ensue." (Stones Justice Manual, 1999: 8-23567). A case in point is that of *R v Selvage (and another [1982] All ER 96).* At the time, Selvage was an employee at the Driver and Vehicle Licensing Authority (DVLA) (who are responsible for recording driving licence endorsements), and, while there, she attempted to (unlawfully) remove four endorsements from an associate's driving licence. Although there were no actual legal proceedings, in progress, at the time (or even pending), had this person gone on to commit any further motoring offences, then the sentence would have been less than they deserve, because it would not have taken these convictions into account. Arguably, a course of justice had already begun (as evidenced by these endorsements), and so the trial judge ruled that this was an attempt to pervert the course of justice. However, the conviction was quashed on appeal because there were no actual legal proceedings, in progress, *at the time*. Under English law, a person can apply to have certain endorsements lawfully removed from their driving licence, after a period of four (and sometimes 11) years (after the offence). Therefore, the premature removal these endorsements would not *necessarily* pervert the course of justice. It would only affect the course of justice, *if* the person committed an offence, *before* the end of the period allowed for the *lawful* removal of such endorsements. As the Court of Appeal made clear, the DVLA merely *registers* these endorsements, it is not part of judicial system *per se*.

The kind of acts, for which the courts are likely to bring a prosecution (for perverting, or attempting to pervert, the course of justice), fall into three categories: making false accusations, 'loosing' or 'tampering with' evidence, and attempting to 'influence' a witnesses (or witnesses). The latter two categories are the only ones worth looking at in any depth, but, for examples of the former, see *R v Rowell [1978] 1 All ER 665,* and *R v Clegg [1975] 1 All ER 334*, re Pte Aindow.

Losing or Tampering With Evidence

R v Murray [1982] 2 All ER 224.

Murray was required to give a specimen of blood for the purposes of determining whether or not he had been driving under the influence of alcohol. He duly complied. The specimen, that was taken at the police station, was found to contain 157mg of alcohol per 100ml of blood. However, the specimen, that Murray submitted for his own defence, was found to contain only 47mg of alcohol per 100ml of blood! Murray was convicted of attempting to pervert the course of justice. (See also *R v Rafique and Others [1993] 4 All ER 1.*)

R v Sinha [1995] Crim L R 68.

Sinha was a junior doctor, who prescribed a drug containing beta blockers to a patient with asthma. The patient died (the next day) from an asthmatic attack, brought on by the drug. (Sinha had not consulted the patient's medical records before prescribing the medication.) After the patient's death, Sinha removed a number of references to the (asthmatic) condition, from her (the patient's) *computerised* medical records. When the coroner asked to see the patient's written medical records, they could not be found (and so the coroner was given a copy of the *computerised* records instead). Sinha later admitted to altering the patient's medical records, but said that he had done so out of fear of being ejected from the practice (by his senior partners). Sinha was convicted of attempting to pervert the course of justice (by attempting to mislead the coroner's court).

Attempting to Influence Witnesses

R v Andrews [1973] 1 All ER 857.

Andrews was a witness to a motoring offence, the principle suspect for which was a shop-keeper by the name of Reeves. Upon being asked to give a statement (by the police), Andrews sought out Reeves, and intimated that he (Reeves) was in serious trouble. Andrews then told Reeves that (for a price) he could furnish him with an alibi. The police overheard this, and Andrews was convicted of attempting to pervert the course of justice.

R v Kellett [1975] 3 All ER 468.

Kellett did not have a very good relationship with either his wife or his neighbours. When his wife began divorce proceedings against him, she asked their (his) neighbours to give evidence on her behalf. Aware of this, Kellett asked a friend to pose as a prospective tenant (of Kellett's), and surreptitiously recorded what the neighbours said about him. Kellett then threatened to sue these neighbours for slander, but made it clear that he would be willing to consider withdrawing his action, if they withdrew the statements that they made (about him), on behalf of his wife. Kellett was convicted of attempting to pervert the course of justice.

R v Toney [1999] 2 All ER 409.

Toney's brother was due to be tried on a charge of robbery. The main witness in the trial was a man named May. Just before the trial, Toney approached May (who he was at school with) and told him that his brother had not been at the scene of the crime. Toney was convicted of attempting to pervert the course of justice. (See also *R v Panayiotou and Another [1973] 3 All ER 112.*)

Using a False Instrument

Under the Forgery and Counterfeiting Act 1981:

> *It is an offence for a person to use an instrument which is, and which he knows to be, false, with the intention of inducing somebody to accept it as genuine, and, by reason of so accepting it to do or not to do some act to his own or any other person's prejudice.*
>
> ***(s.3 Forgery and Counterfeiting Act 1981)***

For the purposes of the Act, an instrument is defined (among other things) as:

> *a) any document, whether of formal or informal character.*
>
> ***(s.8 Forgery and Counterfeiting Act 1981)***

For the purposes of the Act, an instrument is defined as false:

> *a) if it purports to have been made in the form in which it is made by a person who did not in fact make it in that form; or*
>
> *b) if it purports to have been made in the form in which it is made on the authority of a person who did not in fact authorise its making in that form; or*
>
> *c) if it purports to have been made in the terms in which it is made by a person who did not in fact make it in those terms; or*
>
> *d) if it purports to have been in the terms in which it is made on the authority of a person who did not in fact authorise its making in those terms; or*
>
> *e) if it purports to have been altered in any respect by a person who did not in fact alter it in that respect; or*
>
> *f) if it purports to have been altered in any respect on the authority of a person who did not in fact authorise the alteration in that respect; or*
>
> *g) if it purports to have been made or altered on a date on which, or at a place at which, or otherwise in circumstances in which, it was not in fact made or altered;*
>
> *or*

> *h) if it purports to have been made or altered by an existing person but he did not in fact exist.*
>
> ***(s.9 Forgery and Counterfeiting Act 1981)***

This offence, therefore, covers every kind of 'document' (even 'informal' ones), and ever permutation of creating, authorising, or altering the same, for the purposes of inducing someone to act in a way that they would not have done so, had the false document not existed. This latter criteria seems to be necessary, in order to secure a conviction (see, for example, *R v Tobierre [1986] 1 All ER 346*).

Perjury

Perjury has always been seen as a very serious offence. In Anglo-Saxon times, the penalty for perjury was the severing of a hand (the *Council of Justice*, 1973:5.) Even today, the offence carries a sentence of up to seven year's imprisonment (s.1, Perjury Act 1911). While the Christian faith is attributed with giving perjury its religious significance ('the judges shall make diligent inquisition: and behold, if the witness be a false witness and has testified falsely against his brother, then shall ye do unto him as he had thought to have done unto his brother.'(Deuteronomy, 19:18, in the *Council of Justice*, 1973:4), this does not explain why the offence has always been punishable with such harsh penalties (see the *Council of Justice*, 1973, for a brief history), even in pagan (Anglo-Saxon) times. Arguably, the reason why perjury has always been punished so severely is because it undermines the legal process. (It is a crime against justice itself.) While the offence has been recognised throughout history (ibid.), the elements of the modern definition have their origins in *R v Aylett (1785) 1 TR 63, 69* (ibid: 1973:7). In his summing-up of this case, Lord Mansfield offered us the following test:

> I take the circumstances [for perjury] necessary to be these: the oath must be taken in a judicial proceeding, before a competent jurisdiction, and it must be material to the question depending.
>
> ***(ibid: 1973:7)***

This is, *essentially*, the definition that the courts use today. The contemporary definition (as found in the Perjury Act 1911) being:

> If any person lawfully sworn as a witness or as an interpreter in a judicial proceeding wilfully makes a statement material in that proceeding, which he knows to be false or does not believe to be true, he shall be guilty of perjury.
>
> ***(s.1, Perjury Act, 1911)***

In order to secure a conviction, a prosecutor must, therefore, show three things: 1) that the statement was *false*, 2) that it was made *wilfully*, and 3) that it was *material* to the outcome of the proceedings. In most cases, the false statement is a deliberate lie, thus satisfying (at least) the first two criteria. For example:

R v Peach [1990] 2 All ER 966.

Two brothers found some Celtic relics, near a local church. In order to prevent the relics from being declared treasure trove (and thus retained by the crown), they told the court that they had found them under the floorboards of one of their houses. One of the brothers later admitted (to a couple of local archaeologists) that they had actually found them near the church. The two brothers were convicted of perjury.

R v Hudson R v Taylor [1971] 2 All ER 244.
Hudson and Taylor (two young women, aged 17 and 19 respectively) were (prosecution) witnesses in an assault case. Just before the trial, Hudson was approached by a group of men 'with a reputation for violence', who threatened to 'cut her up', if she 'told on' the accused. She told Taylor, and as a result, both women failed to identify the accused in court. The two women were charged with perjury. They admitted to the offence, but appealed on the grounds of duress. Their appeal was allowed. (See also *R v Mullany (1865) 13 WR 726,* and *R v Leigh and Harrison [1966] 1 All ER 686.*)

In the above cases, the false statement was clearly material to the outcome of the trial. However, there are cases where the statement is clearly false, but there is some doubt as to its materiality. To this end (determining materiality), the courts seem to have adopted a number of 'rules' [1]:

Does the false statement enhance the believability of other (possibly false) statements?

R v Tyson (1808) 16 WR 317.
Tyson was a witness (read alibi) for Sullivan, who was accused of robbery. At the trial, Tyson testified a) that Sullivan had been at a certain house, on the night in question; b) that he (Sullivan) had lodged there for the past two years; and c) that he (Sullivan) had never left he house for more than three nights in a row. Only the last two statements were (proven to be?) false, but it was held that these statements have a tendency to enhance the believability of the first (which makes them material), which made Tyson guilty of perjury.

Director of Public Prosecutions (DPP) v Humphreys [1976] 2 All ER 497.
Humphreys was charged (on 18 July 1972) with driving a motor cycle while disqualified. He denied that he was the person riding the motor cycle, on that day, and said (in court) that he had not ridden his motor cycle *at all* during that year (1972). Humphreys was acquitted, but the police found witnesses who said that they had seen him riding his motor cycle that year (1972). Humphreys could not be re-tried for the original offence, but he was found guilty of perjury, because the statement he made about not riding his motor cycle that year enhanced the believability of his other statement about riding his motor cycle on the day of the offence. He was, therefore, found guilty of perjury.

Does the false statement effect the credibility of the person as a witness?

R v Gibbons (1862) 10 WR 1111.
In a paternity case, one of the witnesses, Ms Bishop, was asked whether she and (a) Mr Harmer had had sexual relations, during the month of September 1860. Ms Bishop testified that they had not. However, Mr Harmer testified that they had! Although this was not material to determining the paternity of the child (the child was born in March 1861, which

means that Mr Harmer was clearly not the father), it was held that Ms Bishop's statement was material to her credibility as a witness (for subsequent proceedings), and so she was found guilty of perjury.

R v Millward [1985] 1 All ER 860.
Millward was a police officer who had cause to stop a car driven by two Indian gentlemen. The driver failed to produce his driving licence, which meant that he would have to produce it at his local police station (within two weeks). However, the licence that was produced, at the police station, was not that of the driver, but the man who was the passenger at the time. The police officer knew the men by sight (i.e. he knew which of them had been driving), and they were charged with deception. While they were waiting to go into the court, one of Millward's colleagues (another police officer, by the name of Revelle) approached the men, and asked to see their driving licences. This was done in the presence of Millward, and it followed a conversation between Millward and the other officer. The two Indian men reported this to their attorney, and Millward was questioned about it in court:

Attorney: Have you spoken to PC Revelle this morning?
Millward: *Yes* (true).
Attorney: Are you aware that PC Revelle approached my clients this morning and asked to see their driving licences...?
Millward: *No* (a lie).
Attorney: Did you ask PC Revelle to see my clients?
Millward: *No* (another lie).
Attorney: Did PC Revelle talk to you after he had spoken to my clients?
Millward: *No* (another lie).
Attorney: Is it not the case that Pc Revelle to see my clients so that you could find out which one was...and which one was...?
Millward: *No* (another lie).
Attorney: Has there been any sort of conversation between you and PC Revelle even remotely resembling that suggested to you?
Millward: *No* (another lie).

The matter was investigated, and Millward was shown to have been lying. This led to the two Indian men being acquitted. This was clearly material to Millward's credibility as a witness, and so he was found guilty of perjury. (See also *R v Holden (1872) 12 Cox CC 166,* and *R v Sweet-Escott (1970) 55 Cr App Rep 316.*)

However, not all (false) statements are relevant to the outcome of the case:

Hewitt (1913) 9 Cr App Rep CCA 192.
Mrs Richards (a shop-keeper) was charged with selling margarine as butter. Her supplier was Mr Hewitt, who, in her *defence*, made the following statements: "I bought no butter in Wellingborough or margarine that I know of on Wednesday, the 20th August." "I cannot account for the butter I supplied turning out to be margarine." Mrs Richards admitted that she had sold the margarine as butter, and so her defence of 'good faith' (and Mr Hewitt's

statement) was immaterial to the outcome of the trial. (See also *R v Wheeler [1916-17] All ER 1111.*)

Similarly, not all statements *are* 'false':

R v Hadfield (1886) 16 Cox CC 148
Hadfield was a health inspector who had determined that a number of butchered rabbits were unfit for human consumption. He immediately seized the rabbits, and proceeded to make a detailed inspection of each one. Under cross-examination, he testified: "I examined all the forty-six. I took them all in my hand. Grayson [the vendor] did not ask me to point out what was wrong with them. Mr Parish was brought on to the scene. I did not keep my gloves on. I took all the forty-six rabbits up to examine separately." Although this can be taken to imply that he examined all the rabbits, individually, before seizing them, he does not actually say this, and so it is not, therefore, a false statement.

Similarly, in the O. J. Simpson trial (*Court TV's* original coverage), one of the pathologists (Mr Fung) said that he had placed a container of evidence into a particular vehicle. When one of O. J. Simpson's attorneys produced a video showing that he had not *personally* place the container into the vehicle, and asked *'Where is it Mr Fung?'*, the pathologist defended his statement by saying that the term 'he' referred to the actions of him and his staff. This explanation was accepted by the judge.

Moral Issues

Lord Archer is not just on trial for breaking the law; he is on trial for attempting to corrupt the very process itself. These offences have particularly strong moral connotations. However, the questions surrounding Lord Archer's morality are not just restricted to those acts for which he is on trial. This case also call into question Lord Archer's fidelity, and whether he can be said to have honoured his wedding vows. While adultery may not be a crime, it is certainly contrary to one of the ten commandments, and it can be cited as grounds for divorce. Lord Archer's loyalty to his family, friends, lovers, and employees (and theirs to him) is something else that the case will also 'test'. As well as putting his relationship with each of these 'on trial', they will also be called to testify (or support) him. The crimes of which Lord Archer stands accused are not 'out of character' *for him*, they are part of continuum of behaviours, characterised by the attitude that the rules, by which everyone else has to live, do not apply to him.

The Case (Study)

Background

This case effectively began with Lord Archer's action against the *Daily Star*. In (late October) 1986, an article appeared in the *Daily Star*, claiming that Lord Archer had slept with a prostitute (Monica Coghlan, or 'Deb'/'Debbie' as she is known to her clients), when 'in fact' Lord Archer was having drinks with a friend (the late Terence Baker) (*News of the World*, 21 November 1999). On the evening in question (8 September

1986), Lord Archer had diner (at Le Caprice) with his literary editor (Richard Cohen, accompanied by Mrs Cohen), until around 10:30, when he joined a friend (Terence Baker) in the bar. They remained in the bar until around midnight, when Lord Archer drove Mr Baker home (*News of the World*, 21 November 1999). However, Ms Coghlan claims (claimed) that she was (having sex) 'with' Lord Archer, at the Albion Hotel, until around 12:45 am, on the morning of 9 September 1986. Ms Coghlan's version of events conflicts with that of Aziz Kurtha (a former television presenter), who claims to have seen Lord Archer waiting for Ms Coghlan at this time. Six weeks later (on 24 October 1986), one of Lord Archer's friends (Michael Stacpoole) met Ms Coghlan, at London's Victoria Station, and gave her £2,000 in cash. This followed a telephone conversation (between Ms Coghlan and Lord Archer) in which Ms Coghlan claims that a man saw her and Lord Archer together, and was 'pressuring' her. (The money was to enable her to leave the country and lead a 'normal' life.) (*News of the World*, 21 November 1999; Raphael, 1989.)

Also around this time, another of Lord Archer's friends (Ted Francis) wrote a letter to Lord Mishcon (Lord Archer's solicitor at the time) saying that he was dining with Lord Archer (at the Sambuca), the evening of 9 September 1986. According to the *News of the World*, Lord Archer was actually with his former personal assistant (and mistress), Andrina 'Andy' Colquhoun, that evening (*News of the World*, 21 November 1999). Clearly, Ted Francis was protecting Lord Archer, but he was doing so in order to save his marriage, not to help him win his libel case. After all, it was the events of the evening of 8 September / morning of 9 September; not the events of the evening of 9 September / morning of 10 September, that were relevant to the (libel) case. Francis' letter (alibi), the circumstances surrounding its making, and the entry made in Lord Archer's diary, referring to their meeting of 9 September 1986 (*News of the World*, 28 November 1999), are the events that have given rise to the current case, in which Lord Archer stands accused of perverting the course of justice, perjury, and using a false instrument. Clearly, this ill-fitting collection of 'facts' begs some kind of explanation, and it is this explanation that the judicial system is seeking (in current case).

Defining 'The Case'

The trial, and the events leading up to it, can be divided into a number of sub-events: the circumstances giving rise to the libel trial, the libel trial itself, the circumstances giving rise to the perjury trial, the perjury trial itself, and so on. In addition to these (sub)-events, there are two events, which are unrelated to the trial, but have found their way into 'the story'. The first is the opening (and closure) of Lord Archer's play, *'The Accused'*, in which he invites (invited) the audience to judge the guilt or innocence of his character (see, for example, *Independent*, 2 March 2000; *Independent*, 27 September 2000; *Sun*, 27 September 2000; *Independent*, 30 September 2000; *Sun*, 30 September 2000; *Independent*, 20 October 2000; *Independent Sunday*, 26 November 2000; *Sunday Telegraph*, 3 December 2000; *Independent*, 6, 7 and 9 December 2000; *Mirror*, 15 December 2000; *Sunday Telegraph*, 17 December 2000; *Independent*, *Mirror*, and *Sun*, 20 December 2000; and *Sunday Times*, 21 January 2001). The second is the death of Monica Coghlan: who was killed by a robber fleeing the scene of his crime (see, for example, *Daily Mail*, *Guardian*, *Independent*, *Mirror*, and *Sun* 28 April 2001; *Mail on*

Sunday, *People*, and *Sunday Mirror*, 29 April 2001; *Independent*, 30 April 2001; *Sun* 30 April 2001; *Independent*, *Mirror*, and *Sun*, 2 May 2001; *Independent*, *Mirror*, and *Sun*, 16 May 2001; *Daily Express*, *Daily Telegraph*, *Guardian*, *Independent*, *Mirror*, *Sun*, and *Times*, 5 July 2001; *Daily Express*, *Daily Telegraph*, *Mirror*, *Sun*, and *Times*, 7 July 2001). These two, rather ironic (and perhaps tragic) events, did not go unnoticed by the press.

Clearly, the events of this case extend beyond the trial itself. It is, therefore, important to define what constitutes 'the case (study)', for the purposes of this investigation. The original (libel) case is only relevant in so far as it is part of the circumstances leading up to the current (perjury) case. Similarly, Ms Coghlan's (life and) death, and Lord Archer's play, are only events that happened to people connected with the trial. They are not about the trial itself. Clearly, the trial itself is relevant, which leaves just the events leading up to the trial. One of these events was a conversation between Archer and Francis, in which they discuss Francis' letter to Mishcon, and the circumstances surrounding its making. It is this conversation, which gave rise to Lord Archer being charged with perverting the course of justice, perjury, and using a false instrument, and so forms part of 'the case'. (The conversation was recorded, transcribed, printed, and made available on a dedicated telephone line, by the *News of the World*, and is hereafter referred to as the *News of the World* tapes.) For the purposes of this investigation, it will be deemed to mark the beginning of 'the case (study)'.

Events Giving Rise to the Perjury Trial

The following is a transcript of the telephone conversation(s), which took place between Lord Archer and Ted Francis, on or around 4 November 1999 (*News of the World*, 28 November 1999).

Francis: ....[to Archer] Have you had Michael Crick (Archer's unauthorised biographer....) on the phone to you?...[H]e's cracking on about the letter I wrote to Mishcon (Archer's Lawyer)....on 22nd January [1987].... It's the one about the alibi.....

Archer: Yeah, well don't ever let him see it. It's the wrong date though, isn't it?.....

Francis: Well, yeah.... We had dinner there didn't we?.....Sambuca....you said to me: 'What were you doing on September the ninth last year?' And I said: 'B** if I know.' You said: 'We had dinner here, didn't we?' And I said: 'Well, if you say so.' Do you remember?

Archer: Yeah, well we've got to be careful, Ted. We don't want to go to a court of law with this.....

Francis: [H]ow the hell could he have found out?....You see what I'm worried about? I don't want to see it [the letter?] in the public domain....And most of all I don't want my family to find out that I told a porkie [a derivative of the cockney rhyming slang for a lie - 'pork pie': lie; hence porkies], because you know what it's like with my family.

Archer: Well, there's no proof this is a porkie....In my diary, it's 15 years ago.

Francis: Have you got your diary for that?....Is my name in it?....

Archer: You're down there. Dinner with me.

Francis: On September the ninth.

Archer: Yup.....

[Francis telephoned Archer again on the 9 November 1999.]

Francis: ....Crick is saying that he knows for certain that we cooked up that alibi....He has given me the evidence, or he has told me about the evidence.

Archer: And what is the evidence?

Francis: He says that I was at the Grand Hotel in Brighton and that he has got a credit card, one of my credit card receipts for it.....

Archer: But that night isn't relevant.....The night that we were meant to have been at that restaurant....Was not the night that I am meant to have seen the girl....They are not even connected....Didn't you realise that?....

Francis: So why did we cook up the alibi?

Archer: Because they changed the day. And then they changed it back.

Francis:: And you were covering?

Archer: Yes....I was having dinner in the restaurant that night....It was with someone else and that was the only reason.....

Francis: Oh, I needn't have written to the lawyer.
Archer: No, you needn't have written at all. Absolutely irrelevant.....It is totally and completely irrelevant.

Francis: OK. Thank [G]od for that....
[Archer telephoned Francis, later that evening.]

Archer: ....[to Francis] There is a good side and a bad side to this. The good side is that I was at Sambuca, that's the good side, i.e. my alibi, I was at Sambuca, is fine. The fact that I got it wrong who I was with is incidental in a way, if you think about it. Because it is where I was that matters, not who I was with.....I was at the restaurant but not with someone I would want the world too know I was with. I was with er...

Francis: I know…

Archer: …who was nothing to do with the business at all but I didn't need it out in the open

Francis: Of course. It is a much more personal affair.

Archer: Absolutely. Much more…..

Francis: I mean the only thing I am worried about is that can he (Crick) prove that you and I actually colluded over that alibi?

Archer: No…you just got it wrong….I was at the restaurant and you just got it wrong…..I am very happy to tell you, it is in the restaurant book, it is in my diary.

Francis: Does it have her name in there?….

Archer: No….

Francis: Does the person you were with know about our arrangement?

Archer: No….

[After talking to his solicitor, Archer telephones Francis again on 12 November 1999.]

Archer:….I thought the wisest thing to do was to have a chat to Mishcon….he says as long as you say nothing they have got nothing.

If *mens rea* (having a guilty mind) was an offence in itself, then this conversation must surely be enough to convict Ted Francis. (It is even clearer in the full transcript.) Francis is clearly worried about something, in spite of Archer's reassurances that his alibi did not feature in the original case. Although Lord Archer is clearly trying to reassure his friend, this does mean to say that his point (about the alibi not being relevant) is not a valid one. This aside, it is interesting to see how Lord Archer changes his story throughout the course of this conversation. Firstly, he tells Francis that he is named (in his diary) as the person he was having dinner with that evening (9 September 1986). Later, he tells him that his diary contains no reference as to *who* he was with that evening. Similarly, Lord Archer starts off by saying that he needed an alibi for 9 September (1986), because this was the day that the *Star's* QC *(initially)* thought he was with Monica Coghlan. Later, he tells Francis that the alibi was to protect the person, who he was 'actually' with that night. (Again, this can be seen more clearly in the full transcript.) One wonders why Lord Archer keeps changing his story, and if there is not more to this than meets the eye.

The above conversation is interesting in light of the questions raised (in the original trial) about Lord Archer's alibi for the evening of 8 September 1986 (as 'reported' in *Raphael*, 1989):

Archer told the jury he began to table-hop before leaving around midnight with his theatrical agent, Terence Baker.

Baker in his evidence supported this, saying he was driven back to his home in Camberwell by Archer in his light grey Daimler, getting home at 1.15am. Under cross examination Baker was asked how he could be so sure that he had met Archer at the Caprice on that particular night as he had no record of a bill, no cheque stub, no diary entry or credit card voucher to prove it. He replied 'I just knew'. Michael Hill [attorney for the Star newspaper] pressed him on this crucial point:

Hill: I must ask you again and I ask you to concentrate on the question: Why did you accept that it was September 8th that you had seen Mr Archer? You have no basis for recalling it yourself so far as I can see.
Baker: I had not seen him…for a very long time. I remember very clearly, he is a distinguished client of mine and if you don't see somebody for a long time you tend to remember very clearly when you do meet them.
Hill: When did you next see Mr Archer?
Baker: I cannot honestly remember….a few months later I think, a few weeks later.
Hill: Before or after this story broke?
Baker: Well, after….

….Archer claimed to the court that he first contacted Baker to see if he would confirm his presence two weeks after the *News of the World's* story on his agents return from holiday around about November 15th . Baker gave a different account telling the court that Archer had contacted him immediately after the *News of the World's* story appeared on October 26th .

The implication that Michael Hill repeatedly put to the jury was that Archer had had considerable in constructing an alibi and was unsure in the immediate aftermath of the News of the World story what to say about his whereabouts that night….In the course of my research for this book, I (Raphael) came across a diary entry kept by Archer's political secretary David Faber which indicated that the Tory Deputy Chairman (Archer) had arranged to dine with his key alibi witness on Tuesday September 9th .

(Raphael, 1989:24-7)

With this in mind, let us now look at the events of the trial (as reported in the English national press).

Summary and Topic Chronology of the Trial

Archer and Francis' first court appearance was on 12 December 2000. The purpose of this was to see how they pleaded to the charges put before them (see above for details), and, if necessary, to set a date for the trial. Both men pleaded not guilty, and were bailed to appear, in court, again on 14 May 2001 (*Guardian*, *Independent*, *Mirror*, and *Sun*, 13 December 2000). Although the trial was not due to begin until 30 May 2001, a few weeks ahead of this date (the week commencing 4 May 2001), Lord Archer was given an opportunity to hear the arguments that would be put at his trial (*Guardian*, 10 May 2001; *Independent Sunday*, 13 May 2001; *Sunday Telegraph*, 13 May 2001).

The trial began, on 30 May 2001, with the Crown's QC (David Waters) painting a picture of the meeting between Archer and Francis (at the Sambuca): "At dinner, Jeffrey Archer asked Ted Francis: 'What were you doing on September 9 1986?'....As one might expect, Ted Francis said he didn't know. Jeffrey Archer asked: 'Could it be possibly you were having dinner with me here – will you do that?' Ted Francis agreed...There followed the subterfuge of Jeffrey Archer introducing Ted Francis to staff at the restaurant, so that, as Jeffrey Archer put it, 'they will remember you being with me but not the date'....[and finally] Jeffrey Archer said: 'About the money you wanted, how much was it?' '£20,000,' said Mr Francis. 'OK, it's as good as yours,' said Jeffrey Archer." (*Daily Express*, *Daily Mail*, *Daily Telegraph*, *Guardian*, *Mirror*, and *Times*, 31 May 2001, reporting may vary slightly).

First Day of Trial:	*Daily Express*	31 May 2001*
	Daily Mail	31 May 2001*
	Daily Telegraph	31 May 2001*
	Guardian	31 May 2001*
	Independent	31 May 2001
	Mirror	31 May 2001*
	Sun	31 May 2001
	Times	31 May 2001*
	Sunday Telegraph	3 June 2001

(* Includes opening statement by the Crown's QC [2].)

Precise Charges:	*Daily Telegraph*	31 May 2001
	Times	31 May 2001
	Daily Mail	1 June 2001

The prosecution's first witness was Angela Peppiatt (Lord Archer's former personal assistant). Ms Peppiatt began by testifying that Lord Archer asked her to falsify a number of diary entries for the 8, 9, and 10 September 1986. At this point, the jury was read a transcript of the original libel trial (presumably relating to these dates). Under cross-examination, Ms Peppiatt revealed that it was standard practice for Lord Archer to leave her with a book of signed, blank cheques (which she kept in a picnic basket), in order to enable her to pay off Archer's office expenses (including her own). This practice was clearly open to abuse, and it was long before these inferences became outright accusations. During his cross-examination, Nicholas Purnell (Lord Archer's QC) openly accused Ms Peppiatt of fiddling her expenses. (This would, of course, be material to Ms Peppiatt's credibility a witness.) Ms Peppiatt later admitted using some of the blank cheques in order to buy goods for her and her family (*Daily Express*, *Daily Mail*, *Guardian*, *Independent*, *Mirror*, *Sun*, and *Times*, 8 June 2001). Ms Peppiatt also testified that she had bought gifts for Lord Archer's lovers (principally 'Andy' - Andrina Colquhoun), on *her* credit card.

(Thus enabling Lord Archer to hide the affairs from his wife and professional colleagues.) This revelation was followed by a show of support from Mary Archer (joining her husband, in court, the next day).

Prosecution Witnesses:

Peppiatt, False Diary Evidence:	*Daily Express*	1 June 2001
	Daily Mail	1 June 2001
	Daily Telegraph	1 June 2001
	Guardian	1 June 2001
	Independent	1 June 2001
	Mirror	1 June 2001*
	Sun	1 June 2001
	Times	1 June 2001*
	News of the World	3 June 2001

(*Includes something about the News of the World Tapes.)

Transcripts of 1987 Trial Read to Jury:	*Daily Mail*	2 June 2001
	Daily Telegraph	2 June 2001
	Guardian	2 June 2001
	Independent	2 June 2001
	Mirror	2 June 2001
	Sun	2 June 2001
	Times	2 June 2001
Peppiatt's Evidence Resumes:	*Daily Express*	5-8 June 2001
	Daily Mail	5-8 June 2001
	Daily Telegraph	5-8 June 2001
	Guardian	5-8 June 2001
	Independent	5-8 June 2001
	Mirror	5-8 June 2001
	Sun	5-8 June 2001
	Times	5-8 June 2001
Mary Archer's Show of Support (No1):	*Daily Express*	9 June 2001
	Daily Mail	9 June 2001
	Daily Telegraph	9 June 2001
	Mirror	9 June 2001
	Sun	9 June 2001
	Times	9 June 2001

Ms Peppiatt then went on to testify that Lord Archer had had numerous affairs (six or seven), and that, at the time of the libel trial, he and Lady Archer were effectively living separate lives. According to Ms Peppitt, Lord Archer was effectively living with Andrina Colquhoun, at his London flat. She joked that, when Lady Archer came round, her and the other personal assistants would have to 'hunt', for pictures of Mary Archer, after Lord Archer's lovers had hidden them. Ms Peppiatt also said that Lord Archer had lied to the (then) Prime Minister (Margaret Thatcher), and the (then) chairman of the Conservative Party (Norman Tebbit) about the affairs (*Daily Express*, *Daily Mail*, *Daily Telegraph*, *Guardian*, *Independent*, *Mirror*, *Sun*, and *Times*, 12 June 2001). According to Ms Peppiatt, Lord Archer was not only cheating on his wife, but his mistress (Andrina Colquhoun) as well.

Peppiatt's Evidence Resumes	*Guardian*	12 June 2001
	Mirror	12 June 2001
	Sun	12 June 2001
	Times	12 June 2001
	Daily Express	13 June 2001
	Daily Mail	13 June 2001
	Daily Telegraph	13 June 2001
	Guardian	13 June 2001
	Independent	13 June 2001
	Mirror	13 June 2001
	Sun	13 June 2001
	Times	13 June 2001

Ms Peppiatt also testified that Lord Archer had asked her to find out what television programmes were on, the night of 9 September 1986. In her words:

> "Jeffrey asked me to do some research on the television programmes for the night of Tuesday 9 September between the hours of ten and 12.30."
>
> ***(Peppiatt, reported in the Times, 13 June 2001)***

She then gave this information to Lord Archer, who used it (or so she implied) to alter his statement, which she then typed-up, and gave to his solicitors (Times, 13 June 2001). As 'insurance', Ms Peppiatt asked her junior (Caroline Norman) to counter-sign a memo, confirming that she was acting under instructions from Lord Archer. The exact words of the memo being:

> Under the orders of Jeffrey Archer I filled in the dates from 5th – 12th September as per his Economist diary 1986. The words and times differed from my own diary which was always considered the master book
>
> ***(Pepiatt's Memo, reported in Mirror, 14 June 2001)***

When called to testify, Ms Norman confirmed that she had signed *a* memo, but could not remember the details.

TV Research:	*Daily Express*	13 June 2001
	Daily Telegraph	13 June 2001
	Guardian	13 June 2001
	Independent	13 June 2001
	Mirror	13 June 2001
	Times	13 June 2001

Ms Norman's Testimony:	*News of the World*	10 June 2001
	Daily Express	14 June 2001
	Daily Telegraph	14 June 2001
	Independent	14 June 2001
	Mirror	14 June 2001
	Times	14 June 2001

At this point, the *(Daily) Star's* QC (from the original trial), Michael Hill, was called to testify. It was Mr Hill's testimony that, had he been able to inspect the original diary, and question the relevant individuals, would have unravelled Lord Archer's alibi "like…a damaged sweater" (Hill, 'quoted' in the *Independent*, 15 June 2001, see also *Daily Express, Daily Mail, Daily Telegraph, Guardian, Sun*, and *Times*, 15 June 2001, quotes may vary, *considerably*). This was followed by another show of support from Mary Archer.

Star's QC:	*Daily Express*	15 June 2001*
	Daily Mail	15 June 2001*
	Daily Telegraph	15 June 2001*
	Guardian	15 June 2001*
	Independent	15 June 2001*
	Mirror	15 June 2001*
	Sun	15 June 2001*
	Times	15 June 2001*

(*Includes Mary Archer's show of support, No2.)

Two weeks into the trial, the Crown's QC revealed that the *News of the World* tapes were part of a sting, set up by the well known publicist, Max Clifford. Apparently, Francis's part of the conversation had been scripted by the *News of the World's* crime correspondent, Neville Thurlbeck. Thurlbeck also gave Francis £19,000 for his trouble. (Francis was allegedly offered £150,000, but settled for just enough to enable him to buy a second-hand car, and make a small donation to charity.) Francis only decided to go ahead with the sting, after he discovered that Archer was to be the official Conservative party candidate for Mayor of London. Through the testimonies of Max Clifford and Neville Thurlbeck, Francis was portrayed as a man willing to provide Archer with an alibi for 'the' affair, but not willing to lie for him in court. (Remember, of course, that Francis also stands accused of attempting to pervert the course of justice.) To this end, Thurlbeck also testified that Lord Archer's friends (i.e. Francis) knew about his 'numerous marital indiscretions'. One of these was, of course, Andrina Colquhoun (now Mrs Waddington), who was the next witness to give evidence.

The 'Sting':	*Daily Express*	16 June 2001
	Daily Telegraph	16 June 2001
	Guardian	16 June 2001
	Independent	16 June 2001
	Mirror	16 June 2001
	Sun	16 June 2001
	Times	16 June 2001

Martial Indiscretions:	*Daily Express*	Not Available
	Daily Mail	19 June 2001
	Daily Telegraph	19 June 2001
	Guardian	19 June 2001
	Independent	19 June 2001
	Mirror	19 June 2001
	Sun	19 June 2001
	Times	19 June 2001

Ms Colquhoun was called, principally, to give evidence about her relationship with Lord Archer. Her evidence was given different interpretations by the different newspapers, which, on the face of it, looks as though they are going for a particular 'angle'. However, it may simply reflect the fact that it is genuinely difficult to ascertain the (precise) nature of her relationship with Lord Archer, particularly in the light of what has been said by the foregoing witnesses. In her testimony, Ms Colquhoun denied that she was Lord Archer's 'live-in mistress', and that Lord and Lady Archer were effectively living separate lives, at the time of the libel trial (*Daily Express*, *Daily Mail*, *Independent*, *Mirror*, and Times, 20 June 2001). She also denied receiving any of the expensive gift's that Ms Peppiatt said were bought for her (*Daily Express*, *Daily Mail*, *Daily Telegraph*, *Guardian*, *Independent*, *Sun*, and *Times*, 20 June 2001). When Ms Colquhoun was questioned (by Ted Francis' QC, Roy Amlot) about the possibility of her being with Lord Archer, at the Sambuca, on 9 September 1986, she replied "No, highly unlikely because after 1984 he did not want to be seen with me in a restaurant…we took steps to keep it [the relationship] much more secret than prior to 1984." (*Daily Express*, *Guardian*, and *Sun*, 20 June 2001; see also the *Daily Mail*, *Independent*, and *Mirror* for that day) She also produced copies of her passport, showing that she was in Greece, with her husband-to-be (Robert Waddington), on that day (*Daily Express*, *Guardian*, *Independent*, *Sun*, and *Times*, 20 June 2001). Following this, Francis withdrew his statement that he had been with Lord Archer on that day (*Mirror*, 20 June 2001).

When it came to the question of why the affair ended, Ms Colquhoun told the court that Lord Archer had told her he was looking to further his political career, and needed to 'tidy up' his private life (which meant ending the affair). Many newspapers portrayed this as Lord Archer 'dumping' Ms Colquhoun, for the sake of his political career (*Daily Express*, *Daily Mail*, *Daily Telegraph*, *Guardian*, *Independent*, *Mirror*, *Sun*, and *Times*, 20 June 2001). The *Mirror*'s reporting of this event, actually had very little of Ms Colquhoun testimony in it. The paper chose, instead, to focus on what Francis had said about the affair. Francis' quote: "Mary and I think Margaret Thatcher said he had to give her up", form the basis of the *Mirror*'s headline: "Thatcher 'Told Archer to Dump His Mistress'" (*Mirror*, 20 June 2001). This gave Francis an(other) opportunity to press his point about giving Archer an alibi for the affair, but not for the court case: "I [Francis] said [to Archer] 'Is Mary threatening divorce?' and he [Archer] said 'Well it could get to that….Will you write to Mishcon and say we were having dinner because she [Mary] will believe it coming from you?'…..I [Francis] said if it ever gets to a divorce, don't count on me to back you up in court because there is no way I would ever back you up." (*Mirror*, 20 June 2001, and for partial reports, see the *Daily Express*, *Guardian*, *Independent*, and *Times*, for that day)

Colquhoun's Testimony:	*Daily Express*	20 June 2001
	Daily Mail	20 June 2001
	Daily Telegraph	20 June 2001
	Guardian	20 June 2001
	Independent	20 June 2001
	Mirror	20 June 2001
	Sun	20 June 2001
	Times	20 June 2001

On 20 June 2001, the jury was taken to see the court where Lord Archer's libel trial was held. This was to establish whether or not the *(Daily) Star's* QC (Michael Hill) would have had a clear view of the diary (*Daily Express*, 21 June 2001; *Daily Mail*, 21 June 2001; Independent, 21 June 2001; Times, 21 June 2001). This (effectively) marked the end of the prosecution's case. It was also marked by Lord and Lady Archer, who celebrated by going out for a meal together (*Daily Express*, 22 June 2001; *Daily Mail*, 22 June 2001).

Visit to Libel Court:	*Daily Express*	21 June 2001*
	Daily Mail	21 June 2001*
	Daily Telegraph	21 June 2001
	Independent	21 June 2001*
	Mirror	21 June 2001*
	Sun	21 June 2001
	Times	21 June 2001*

(* Includes legal explanation for the visit.)

The defence's case began with the revelation that Lord Archer would not be taking the stand *himself*. Instead, his QC (Nicholas Purnell) called a number of 'character witnesses' (four, according to the *Independent*, 28 June 2001), to testify as to Lord Archer's good name. These included film producer, Frank Marshall, who testified that Lord Archer had "always been a good friend...kind...generous...sincere and honest" (*Daily Express*, 28 June 2001; Times, 28 June 2001; and for less complete quotes see also the *Daily Mail*, 28 June 2001; *Mirror*, 28 June 2001), and Gilbert Gray QC, who said that Lord Archer was "trusting to others and worthy of trust" (*Mirror*, 28 June 2001)...[a man who wore] "his heart on his sleeve (*Daily Mail*, 28 June 2001; Times, 28 June 2001; although the *Independent* attributes this statement to *Marshall*, 28 June 2001).

Defence Witnesses:

Archer Remaining Silent:	*Daily Express*	28 June 2001*
	Daily Mail	28 June 2001*
	Daily Telegraph	28 June 2001
	Guardian	28 June 2001
	Independent	28 June 2001*
	Mirror	28 June 2001*
	Sun	28 June 2001
	Times	28 June 2001*

	People	1 July 2001

(* Includes character witnesses.)

At this point, Francis' QC (Roy Amlot) produced evidence which (he said) showed that not all the money Lord Archer won in his libel trial (£500,000) went to charity (as Lord Archer had said it would). The receipts from Lord Archer's accounts show that only £130,000 went to charity. Arguably, this is material to Lord Archer's credibility as a witness, or rather the credibility of those witnesses who have testified to his 'good' character. It is not clear what Roy Amlot was looking to achieve by this, but he may have been trying to show that Lord Archer is the kind of man who deceives people (including his friends). This was accompanied by yet another show of support for Lord Archer, this time from his whole family.

Charity Short Changed:	*Daily Express*	29 June 2001*
	Daily Mail	29 June 2001*
	Daily Star	29 June 2001*
	Daily Telegraph	29 June 2001*
	Guardian	29 June 2001*
	Independent	29 June 2001*
	Mirror	29 June 2001*
	Sun	29 June 2001*
	Times	29 June 2001*

(*Includes family show of support.)

The following day, it was Mary Archer turn to give her evidence. Under oath, she denied that her and her husband were effectively leading separate lives, at the time of the libel trial, and produced some of the jewellery which had allegedly been bought for Ms Colquhoun. Lady Archer did, however, concede that a number of outside events had put a strain on the marriage. This she portrayed this as normal, for (a marriage to) someone in Lord Archer's position: "He would not have been the first aspiring politician to have had the odd fling." (*Daily Express*, *Daily Mail*, *Daily Star*, *Daily Telegraph*, *Mirror*, *Sun*, and *Times* 30 June 2001, quotes may vary). She denied 'covering-up' the affair, in order to lend credibility to the suggestion (made by Lord Archer's QC, in the libel) that Lord Archer was not a man to have affairs.

Mary Archer's Testimony:	*Daily Express*	30 June 2001
	Daily Mail	30 June 2001
	Daily Star	30 June 2001
	Daily Telegraph	30 June 2001
	Guardian	30 June 2001
	Independent	30 June 2001
	Mirror	30 June 2001
	Sun	30 June 2001
	Times	30 June 2001

The final witness, for the prosecution, was Ted Francis. His testimony was little more than a reiteration of the circumstances under which the alibi had came about. (As intimated by the Crown's QC, at the start of the trial.) Francis' testimony did, however, include the reason for him deciding to expose Lord Archer. Apparently, Lord Archer had embarrassed him at a party (by announcing to all those assembled: "[t]his man owes me £20,000 and has not paid it back") (Francis, allegedly quoting Archer, *Daily Express*, *Independent*, *Mirror*, 4 July 2001; see also *Daily Star*, 4 July 2001). Unfortunately, Terence Baker (Lord Archer's alibi for the evening 8 September 1986) passed away in the intervening years, which meant that he could not be questioned (*Daily Telegraph*, 7 July 2001; *Independent*, 7 July 2001). Francis' testimony therefore marks the end of the defence's case.

Ted Francis' Testimony:	*Daily Star*	3 July 2001
	Daily Telegraph	3 July 2001
	Guardian	3 July 2001
	Independent	3 July 2001
	Mirror	3 July 2001
	Sun	3 July 2001
	Times	3 July 2001
	Daily Express	4 July 2001*
	Daily Star	4 July 2001
	Daily Telegraph	4 July 2001
	Independent	4 July 2001*
	Mirror	4 July 2001*
	Sun	4 July 2001

(* Includes party jibe quote.)

Terence Baker:	*Daily Telegraph*	7 July 2001
	Independent	7 July 2001

In his summing up, the Crown's QC (David Waters) described Lord Archer as "a man whose instinct and solution, whatever successive allegations or obstacles he faced, was to manipulate events and fabricate a dishonest answer…(*Daily Express*, *Daily Star*, *Daily Telegraph*, *Independent*, *Mirror*, 6 July 2001, quotes may vary)… He is entirely the author of his own misfortune." (*Daily Express*, *Daily Telegraph*, Independent, *Mirror*, and Times, 6 July 2001, quotes may vary) Lord Archer's QC summed-up the case as follows: "Being a bloody fool may be a male prerogative…It may be more an attribute to the male of the species than the female. But being a bloody fool does not make you a criminal. It makes you do things you regret" (*Mirror*, *Sun*, and Times, 7 July 2001, see also *Daily Express*, 7 July 2001, quotes may vary). He is clearly alluding to the fact that the men, in this case, seem to have done some pretty dumb things (paying-off Monica Coghlan, signing blank cheques, having affairs, creating alibis, lying about diaries, taping telephone conversations, and so on), while the women seem to have behaved much more astutely. However, we must not lose sight of the fact that a crime is still a crime, regardless of 'foolishly' the perpetrator is perceived to have behaved. Clearly, both barristers know their job.

In his summing-up, the judge told the jury that they did not have to reach a verdict on the charge of using a false instrument, because this duplicated one of the other charges (*Daily Express*, *Daily Mail*, *Daily Star*, *Daily Telegraph*, *Guardian*, *Independent*, and *Times*, 12

July 2001). On the matter of Lord Archer's silence, Mr Justice Potts (the judge) directed the jury to see Lord Archer's silence, not as evidence of guilt, *in itself*, but as evidence that may support the prosecution's case. In other words, could Lord Archer not defend himself? With regard to the diary, Mr Justice Potts directed the jury as follows: "You must ask yourselves what the consequences would have been in Lord Archer's libel trial, if those pages in that genuine diary had been available to the judge and jury." (*Daily Mail*, *Daily Star*, *Daily Telegraph*, *Independent*, and *Times*, 12 July 2001, and the *Mirror*, 12 July 2001, for a less complete quote – quotes may vary) He is clearly directing the jury to see those statements, made about the diary, as material to the outcome of the libel trial

Summing Up:

Crown's QC Summing-up:	*Daily Express*	6 July 2001
	Daily Star	6 July 2001
	Daily Telegraph	6 July 2001
	Independent	6 July 2001
	Mirror	6 July 2001
	Sun	6 July 2001
	Times	6 July 2001
Archer's QC Summing-up	*Daily Express*	7 July 2001
	Daily Star	7 July 2001
	Mirror	7 July 2001
	Sun	7 July 2001
	Times	7 July 2001
	Daily Star	11 July 2001
	Guardian	11 July 2001
	Independent	11 July 2001
	Mirror	11 July 2001
Judge's Summing-up:	*Daily Express*	12-? July 2001
	Daily Mail	12-13 July 2001
	Daily Star	12-13 July 2001
	Daily Telegraph	12-13 July 2001
	Guardian	12-13 July 2001
	Independent	12-13 July 2001
	Mirror	12-13 July 2001
	Times	12-13 July 2001
	Sun	13 July 2001

After five days of deliberation, the jury finally reached its verdict (on 19 July 2001). They found Lord Archer guilty on two counts of perverting the course of justice, and two counts of perjury (for which he was sentenced to four years imprisonment), and they found Ted Francis not guilty of perverting the course of justice. Following the verdict, Lord Archer's QC (Nicholas Purnell) was quick to make the point that the lies of 14 years ago had not been compounded by the current trial. He was interrupted by Mr Justice Potts, who asked: "What about the evidence of Lady Archer?" (*Daily Express*, *Daily Mail*, *Daily Telegraph*,

Mirror, *Sun*, and *Times*, 20 July 2001, quotes may vary) He is referring to the fact that, during the course of the trial, Lady Archer made a (sworn) statement to the effect that the false diary was genuine, and that it had been in existence for some time (prior to the libel trial) (*Daily Express*, *Daily Mail*, *Mirror*, and *Times*, 20 July 2001). This prompted a police investigation to decide whether or not to bring charges against *Lady* Archer (for perjury) (*Daily Express*, *Daily Mail*, *Daily Star*, *Daily Telegraph*, *Guardian*, *Independent*, *Mirror*, *Sun*, and *Times*, 20 July 2001).

Jury's Deliberation:	*Daily Express*	14-19 July 2001*
	Daily Star	14-19 July 2001*
	Daily Telegraph	14-17 July 2001*
	Guardian	14-19 July 2001*
	Independent	14-19 July 2001*
	Mirror	14-19 July 2001*
	Sun	16-19 July 2001*
	Times	14-19 July 2001*

(* Not necessarily on all days.)

The Verdict:	*Daily Express*	20 July 2001
	Daily Mail	20 July 2001
	Daily Star	20 July 2001
	Daily Telegraph	20 July 2001
	Guardian	20 July 2001*
	Independent	20 July 2001*
	Mirror	20 July 2001
	Sun	20 July 2001
	Times	20 July 2001*

(*Includes verdicts on specific charges.)

While this marks the end of the *trial*; it is clearly not the end of the *story* (see English national press, 20 July 2001 onwards). For the purposes of this investigation, however, it marks the end of 'the case' (study).

General Discussion

As the above analysis shows, each day was dominated by a particular witness or event in the day's proceedings. Moreover, there was a high level of agreement, among the different newspapers, as to what that was. Unlike the earlier paper (Taylor, 1997), there was not that much difference between the topics covered by the tabloids and those covered by the broadsheets. There was, however, some marked differences in the reporting of particular topics. For example, the *Times*, *Daily Telegraph* (broadsheets), and *Daily Mail* (tabloid) were the only titles to include the exact wording of the charges, in their reporting of the first day's events. While, *technically*, this is a mix of tabloid and broadsheet coverage, it would be fair to say that the *Daily Mail* does take itself more seriously than the other tabloids. Similarly, the *Guardian*, *Independent* and *Times* (all broadsheets) were the only titles to include the (precise wording of the) charges in their reporting of the verdict. The implications of Terence Baker's death only featured in two titles (*Daily Telegraph* and Independent), both broadsheets. Most titles covered the visit to the libel court, but only the *Daily Express*, *Daily Mail*, *Independent*, and *Times* (former two tabloid, latter two broadsheet) explained the (legal) reasons for this. The Times' coverage of Ms Peppiatt's testimony was far more comprehensive than any other title, but the *Mirror* (tabloid) was the only one to include the wording of the memo. The Archers' evening out was only reported in two of the tabloids (*Daily Express* and *Daily Mail*). Perhaps more telling is the fact that the *Guardian* and the *Independent* (both broadsheets) did not seem to think that Lady Archer's *first* show of support (for her husband) was worth reporting. Although these topics do not break down into a 'neat' broadsheet/tabloid split, the broadsheets do seem to focus *more* on the legal *details*.

These reports are, however, not without their 'biases', and while they are not all *that* much of a problem (for the method), it is still perhaps worth pointing them out. The first is the way in which different journalists/titles use different forms of address. While most titles refer to the defendant as '*Lord* Archer', the *Independent* refers to him as simply '*Jeffrey* Archer'. This may be a thinly disguised show of contempt, or it may be a way of subtly 'slanting' their reports. This may or may not have anything to do with the fact that the *Independent* was the only *broadsheet* to report details of the jibe that he (Lord Archer) made about Ted Francis. The second (and this perhaps follows on from the first) is that Nicholas Purnell's (Lord Archer's QC) summing-up received less press coverage than David Water's (the Crown's QC). Here again, the press may have been sending out a very subtle message. More of a concern is the inaccuracy of the 'quotes': Michael Hill (the *Star's* QC)'s quote, about Lord Archer's alibi unravelling like a damaged sweater, the quotes of Lord Archer's character witnesses, and the quotes of Mary Archer, are all worded differently, in different newspapers. This suggests that a 'quote' may not be what it purports to be (i.e. a report of what that person said *verbatim*), but a record of what the reporter 'remembers' (that person saying) and/or could write down at the time. (For a demonstration of the reconstructive nature of memory, see Sir Frederick Bartlett's (1932) '*Remembering: A Study in Experimental and Social Psychology*.) One would have thought that even if the report is not impartial, at least the 'quotes' would be accurate. This is particularly disappointing, from a researcher's point of view, because a person's actual words can often be very informative. One of the ways round this problem is to sample a number of titles, in order to see whether it is possible to arrive at a 'consensus' of what was said (as has been done here). Even then, there is no guarantee that this is what the person actually said.

When it comes to the legal topics/issue evidenced in the case, clearly there is more going on here than just that which relates to the charges against Archer and Francis. It certainly appears that Ms Peppiatt may have had good reason to distort at least some of her testimony (i.e. to cover up her own abuses of Archer's financial system and her feelings of animosity towards her former employer). It also appears that Ted Francis' testimony should not be taken at face value either. With regard to Mary Archer's testimony, the judge has already drawn our attention to at least one lie (and there may be more). In this case (and perhaps all cases) the lawyers did not just have to present the evidence (for their case), they also had to establish the credibility of the witnesses giving that evidence. They may also have been mindful of charges that could be brought against these witnesses. In order to see what is relevant to what topic/issue, it is necessary to look at them separately.

Evidence and Issues Relevant to the Charges Against Archer

During the course of the trial, the judge (Mr Justice Potts) told the jury that they did not have to come to a decision on the charge of using a false instrument (because this duplicated one of the other charges) (*Daily Express*, *Daily Mail*, *Daily Star*, *Daily Telegraph*, *Guardian*, Independent, and Times, 12 July 2001). This left three charges of perverting the course of justice: (count one) procuring Ted Francis to provide his [Archer's] solicitors with a version of events.....which he knew to be false, and (count two) failing to reveal the existence of his main office diary...., [later providing] Angela Peppiatt with a blank 1986 diary [in] which he instructed her [to make certain] entries [and] handed [this] to his solicitors for use in the court case as his main office diary, (count three) [making] entries in [a]...diary for September 1986 which purported to have been made on or before that date, and two charges of perjury: (count one) [making] a statement under oath for use in judicial proceedings that the [diaries provided in court] were the only documents of that type which had been in his possession, and (count two) as a sworn witness in judicial proceedings...[making] a false statement that the diary known as the main diary...was in existence and contained the entries relating to September 8 and 9 1986, before October 26, 1986, to be decided upon by the jury (*Daily Telegraph*, 31 May 2001; *Times*, 31 May 2001; *Daily Mail*, 1 June 2001; see also the *Sunday Telegraph*, 13 May 2001 for a slightly less precise account of the charges).

The *News of the World* tapes were proof that Archer had procured Francis to provide him with a false alibi (for the night of 9 September 1986). Consequently, he was found guilty on this count of perverting the course of justice (*Guardian*, *Independent*, and *Times*, 20 July 2001). With regard to those charges relating to the diary, the jury was given clear instructions (by the judge) as to how they should look upon this evidence: "You must ask yourselves what the consequences would have been in Lord Archer's libel trial, if those pages in that genuine diary had been available to the judge and jury." (Mr Justice Potts, reported in the *Daily Mail*, *Daily Star*, *Daily Telegraph*, *Independent*, and *Times*, 12 July 2001, and the *Mirror*, 12 July 2001, for a less complete quote – quotes may vary). Clearly, there is no evidence that Lord Archer made any changes to these diaries *himself* (they were all in Ms Peppiatt's handwriting), and so the jury found him not guilty on this charge (*Guardian*, *Independent*, and *Times*, 20 July 2001). With regard to the charge of providing Angela Peppiatt with a 1986 diary, and instructing her to make false entries in the same, the jury was clearly satisfied that this had happened (from Ms Peppiatt's testimony, Ms Norman's testimony, and the documentary evidence supplied by Ms Peppiatt), and that it had perverted the course of justice in the original (libel) trial. It follows, therefore, that the statements Lord Archer made (on oath) about the these diaries (at the time) must have been false (and made wilfully). Moreover, they were seen as being material to this case (as directed by the judge). The jury, therefore, found Lord Archer guilty (on both counts) of perjury (*Guardian*, *Independent*, and *Times*, 20 July 2001).

Evidence and Issues Relevant to the Charge Against Francis

The charge against Francis was one of perverting the course of justice, specifically that he had...provided [Archer's] solicitors [with] a version of events....which he knew to be false and materially connected with an action against *Express Newspapers* (*Daily Telegraph*, 31 May 2001; *Times*, 31 May 2001; *Daily Mail*, 1 June 2001; see also the *Sunday Telegraph*, 13 May 2001 for a slightly less precise account of the charges). Francis was found not guilty on this charge (*Guardian*, *Independent*, and *Times*, 20 July 2001), and it is not difficult to see why; Francis' alibi did not relate to the evening in question, and so he was not called to give evidence (at the libel trial). The letter that he wrote to Lord Mishcon did not, therefore, effect the outcome of the (libel) trial (and so did not pervert the course of justice).

Credibility of Witnesses

From the *News of the World* tapes, it is clear that Ted Francis was lying about being with Lord Archer on the evening 9 September 1986. By his own 'admission' (although this could have simply been part of the script supplied by the *News of the World*), he was at the Grand Hotel, in Brighton, that night. This much we 'know', but what of his reason's for providing Lord Archer with the alibi? Francis appeared keen to make the point that he was willing to provide Lord Archer with an alibi for his affair(s), not for his court case. However, Andrina Colquhoun testified that her affair, with Lord Archer, was virtually over by the end of 1984. She also produced a photo-copy of her passport, showing that she was in Greece, with her husband-to-be, on 9 September 1986. (The day that Lord Archer claims to have been having dinner with her, at the Sambuca.) Is it *conceivable* that Ted Francis could have been close enough to Lord Archer to provide him with an alibi, but not close enough to know that his affair with Ms Colquhoun had ended two years earlier? If Francis' alibi *had* formed part of the original (libel) trial, then perhaps his credibility (as a witness) would have been more of an issue. As it stands, however, it is not *material* to this case.

Angela Peppiatt's credibility was, however, much more of an issue. Although Ms Peppiatt admitted to fiddling her expenses, we still do not know what happened to the jewellery and gifts that, *she says*, were bought for Lord Archer's lovers. Did Ms Peppiatt make up these stories about Lord Archer's lovers in order to mask her own dishonesty; or was her account of Lord Archer's 'love' life accurate (save for the part about the gifts)? According to the press, Lord Archer has had various affairs with different women (*Daily Express*, *Daily Mail*, *Daily Star*, *Daily Telegraph*, Independent, *Mirror*, *Sun*, and Times, 20 July 2001), but has he bought them all expensive gifts? The question is perhaps better cast as: 'Is this how a man (like Lord Archer) treats his mistresses, or is this how a woman (like Ms Peppiatt) imagines (and perhaps would like us to believe) that a man (like Lord Archer) treats his mistresses?' Ms Colquhoun says that *she* did not received any of these expensive gifts (just a few toiletries). When it comes to the question of the Lord Archer's relationship with Ms Colquhoun, do we believe Angela Peppiatt or Andrina Colquhoun? Similarly, when it comes to the question of Lord Archer's fidelity, do we believe Angela Peppiatt (and virtually the whole of Fleet Street) or Mary Archer? With Mary Archer, the issue not so much her credibility as a witness, as it is the charges that may be brought against her for perjury.

Moral Questions

The reporting of the trial was deceptively non-judgmental. With the exception of the occasional article about Lord Archer's 'pulling power' (*Daily Telegraph*, 22 June 2001), Andrina Colquhoun's demeanour (*Times*, 20 June 2001), Lord and Lady Archer's relationship (*Mirror*, 20 June 2001), and the changing role of 'personal assistants (*Independent Sunday*, 10 June 2001, *Daily Mail*, 9 June 2001), the press chose not to print any 'feature' articles about Lord Archer (or anyone connected with him), for the duration of the trial. However, as soon as the verdict was known, every single national, daily newspaper came out with a multiple-page feature (*Daily Express*, eight pages; *Daily Mail*, 14 pages; *Daily Star*, eight pages; *Daily Telegraph*, six pages; *Guardian*, five pages; *Independent*, six pages; *Mirror*, 15 pages; *Sun*, eight pages; *Times*, six pages, 20 July 2001 – note the broadsheet pages are twice as large) about Lord Archer's life. These (multiple-page) feature articles chronicled the full extent of Lord Archer's infidelities, underhand business dealings, personal and family infamies, and the lies he told to friends, family and colleagues. Many titles also carried articles in their 'comment' pages (*Daily Express*, *Daily Telegraph*, *Guardian*, *Independent*, and *Times*, 20 July 2001), as well as cartoons about Lord Archer and/or the trial (*Daily Express*, *Daily Mail*, *Daily Star*, *Daily Telegraph*, *Guardian*, *Mirror*, *Sun*, and *Times*, 20 July 2001). This was not simply the press revelling in yet another scandal involving a public figure; they were putting on display, every aspect of Lord Archer's corrupt and sordid past. Lord Archer had been placed in the stocks, and the whole village was turning out to throw fruit at him. The law has already punished Lord Archer for his crimes; this is his punishment for the way he had treated his wife, his friends, his colleagues, his employees, and women in general; it is an old-fashioned punishment (public humiliation), for a man who has snubbed old-fashioned values.

Methodological Discussion

This paper shows how it is possible to 'piece together' an account of a legal case from a number of different press reports. It must, however, be remembered that this is a particularly well documented case, and some events – the *News of the World* tapes, the pre-trial, court sessions, the wording of the charges, the contents of the memo, the details of the various testimonies, and the (legal) relevance of certain pieces of evidence – are all aspects of the case that were reported on (in detail), almost exclusively, by certain titles. With one or two notable exceptions, each of the above titles made *some* unique contribution to the above account. This aside, the high level of consistency between them (the reports), makes it possible to validate the information in most of the reports. This means that the above account is not only valid, but it can be seen to be valid. For this reason (validating other reports), it is important not to write-off the contribution of any title (unless, of course, one has a good reason for doing so). While the broadsheets give the *impression* that they are more concerned with 'serious' news items, and the tabloids give the *impression* that they are more concerned with 'human interest' stories, these self-styled corporate images do not necessarily reflect the amount of the *information* contained in their news reports. A researcher should not, therefore, allow their perceptions of a particular title to 'colour' what they 'see' in its news reports. Clearly this is an issue for an approach such as this, and certainly something that needs investigating further.

As it was originally conceived, this approach was designed to show that valid information could be derived from what has traditionally been regarded as a rather dubious source of information (see, for example Glasgow University Media Group, 1976, 1980; Lee and Solomon, 1992; Philo, Hewitt, Beharrell, and Davis, 1982). This paper has taken the method one step further, by looking at some of the practical problems associated with using such an approach. While it is assumed (in this, and the previous paper) that the reader intends to use this information for academic research, the way in which the case has been presented here also has educational applications. Unlike law reports, which contain only brief summaries of the relevant facts, accounts such as this not only show the judicial process at work; they also allow the reader to take on the role of a juror/barrister, playing-out the case for themselves. To this end, the paper has presented events as they 'happen', not how they can be seen in hindsight. The *Gale Group* database has the facility to download the results of a search to an e-mail account. This can then be cut and pasted to produce an account of events, or separated into different perspectives/roles. This makes it possible to produce a *reasonably comprehensive* account of events (accepting the deficits of the database, highlighted earlier), with far less effort than has been expended here. Clearly this is something that the reader will, no doubt, wish to explore for themselves (and disseminate in future publications).

References

Principal Journalists:

Daily Express – John Twomey, John Chapman, Virginia Blackburn.
Daily Mail – Paul Harris, Nick Craven, Peter Allen, Neil Sears, Susie Boniface.
Daily Star – John McJannet, David Paul, Michael Booker.
Daily Telegraph – Sue Clough, Sandra Barwick.
Guardian – Paul Kelso, Tania Branigan, Owen Bowcott, Stuart Millar.
Independent – Kim Sengupta, Harvey McGavin, Robert Verkaik, Matthew Beard, Terri Judd.
Mirror – Harry Arnold, Lucy Rock, Gill Swain, Lee Brown.
Sun – Simon Hughes, Antonella Lazzeri, Mike Sullivan.
Times – Tim Reid, Steve Bird, Valerie Grove, Ian Cobain, Andrew Norfolk, Adam Sherwin, Matthew Parris,

Legal Cases and Statutes:

Director of Public Prosecutions (DPP) v Humphreys [1976] 2 All ER 497.

Hewitt (1913) 9 Cr App Rep CCA 192.

R v Andrews [1973] 1 All ER 857.

R v Clegg [1975] 1 All ER 334.

R v Gibbons (1862) 10 WR 1111.

R v Hadfield (1886) 16 Cox CC 148

R v Hudson R v Taylor [1971] 2 All ER 244.

R v Kellett [1975] 3 All ER 468.

R v Leigh and Harrison [1966] 1 All ER 686.

R v Mullany (1865) 13 WR 726.

R v Murray [1982] 2 All ER 224.

R v Panayiotou and Another [1973] 3 All ER 112.

R v Peach [1990] 2 All ER 966.

R v Rafique and Others [1993] 4 All ER 1.

R v Rowell [1978] 1 All ER 665.

R v Selvage and another [1982] All ER 96.

R v Sinha [1995] Crim L R 68.

R v Sweet-Escott (1970) 55 Cr App Rep 316.

R v Tobierre [1986] 1 All ER 346

R v Toney [1999] 2 All ER 409.

R v Tyson (1808) 16 WR 317.

R v Wheeler [1916-17] All ER 1111.

Rantzen v *Mirror* Group Newspapers (1986) Ltd and Others [1993] 4 All ER 975.

Forgery and Counterfeiting Act, 1981.

Perjury Act, 1911.

Stones Justice Manual (1999) London: Butterworth.

Academic References:

Bartlett, F. C. (1950 edition) *Remembering: A Study in Experimental and Social Psychology*. Cambridge: Cambridge University Press.

(The) Council of Justice / Hunter, M. (1973) *False Witness: The Problem of Perjury*. London: Stevens and Sons.

Glasgow University Media Group (1976) *Bad News*. London: Routledge and Kegan Paul.

Glasgow University Media Group (1980) *More Bad News*. London: Routledge and Kegan Paul.

Lee, M. A. and Solomon, N. (1992) *Unreliable Sources: A Guide to Detecting Bias in News Media*. New York: Lyle Stuart.

Philo, G., Hewitt, J., Beharrell, P. and Davis, H. (1982) *Really Bad News*. London: Writers and Readers.

Popper, K. R. (1976) *Conjectures and Refutations: The Growth of Scientific Knowledge*. London: Routledge and Kegan Paul.

Raphael, A. (1989) *My Learned Friends*. London: W H Allen.

Taylor, M. S. (1997) *Discourses of Defamation: A Lesson in Mass media Deviance and Legal Control*. Unpublished M.Phil. dissertation.

Taylor, M. S. (2007) *Interacting with Statistics: The Role of Statistics in the Research Process*. Gt Shelford, Cambridge: Cambridge Academic.

Notes

1: These are not recognised legal 'tests', but they do appear to describe the reasoning that underlies these decisions.

2: To qualify, the report must include all three element of the QC's opening comments: the alibi, the subterfuge, and the loan.

Index

Made in United States
Orlando, FL
05 November 2026

90229836R00153